How To Live o[n]

.95   Simon + S[chuster]

630 Fifth ave

New York N.Y.
             10020

Light on yoga

B.K.S. Iyengar

Schocken Books Inc.

67 Park ave

N.Y. N.Y. 10016

Amateur Beekeeping

EDWARD LLOYD SECHRIST

*Formerly Associate Apiculturist
U.S. Bee Culture Laboratory
Davis, California*

Amateur

WITH
ILLUSTRATIONS
BY
FRANK NORRIS TODD

OLD GREENWICH, CONNECTICUT   1974

Beekeeping

THE DEVIN-ADAIR COMPANY

Copyright 1955 by The Devin-Adair Company. All rights reserved.
No portion of this book may be reproduced in any form,
except by a reviewer who may quote brief portions in a review,
with written permission of the publisher, The Devin-Adair Company,
Old Greenwich, Connecticut 06870

Manufactured in the United States of America

ISBN 0-8159-5001-2

Library of Congress Catalog Card Number 55-11865

Designed by Lewis F. White

Second printing 1971

Third printing 1974

Preface

You are a young person who wants to know something about bees, and who wishes to acquire that knowledge through practical experience with your own hives. Or perhaps you are a grandparent, retired from an active city life, and eager to spend your "declining years" in a rural environment, enjoying some of the lighter tasks and healthful outdoor activity which you missed in earlier days, and you would like to produce your own honey, or have a few bees to pollinate your orchard. Anyway, whoever you are, or wherever you live, you want to know how to keep bees. There are many ways, but in this book I shall point out only the one way I like best, and the method I would advise you to follow. Try it faithfully until you have mastered it. Then, and not until then, try other plans if you want to.

Many beekeepers fail because they attempt to follow the advice of different people, instead of pursuing one well-thought-out plan.

To be successful in any line of work, it is necessary to know *why, how,* and *what;* or, to express it as a philosopher might, to consider *end, cause,* and *effect.* You must know the *end* to be attained (or *why* you are engaging in this activity); plan *how* to attain that end, i.e., study the *means* or the *causes* which will best bring it to fruition; and then work for the *effect* desired, which is *what* you have wanted all along.

There are many other books for beekeepers, excellent ones which cover the subject thoroughly; but they are almost without exception for the commercial beekeeper who wants to earn his living, or to supplement his earnings, by producing large quantities of honey. You, however, want to study bee behavior, to be able to "ride your hobby" among your friends, to supply your table with some good honey, or to have bees for pollination; and you do not want to wade through the larger tomes—at least, not just yet! This small book will tell you why you should keep bees, how to help them work to your best advantage, and what may be achieved by them with your cooperation.

As your hand-book, it will give you all the details necessary for success in learning bee behavior and how to handle a colony for good results. Too often the instruction books allow the beginner to flounder along without knowing enough about how to manage his bees intelligently. This book will help you, step by step, in attaining the results you desire. All the essentials are given, together with the most important facts relating to the nature and the customary behavior of the bees and the colony. A complete treatment of many less important details could not be put into one small volume; but if you wish to know still more when you have mastered what is given here, you may turn to the list of excellent references at the end.

As a result, you will become a good amateur or "back lot" beekeeper, producing honey for your family, and, in good seasons, perhaps enough to share with your neighbors. If you have an orchard of fruit trees, your bees will pollinate the blossoms, thus producing more and better fruit. Perhaps later, if you are a young person, you will become a hobbyist beekeeper, a serious student, able to give demonstrations on how to handle bees, and to make talks on them to rural or small town clubs. You may even have a money-making career as a commercial beekeeper.

"A hobby," said Dr. George E. Cutton, President of Colgate University, "has seven distinguishing characteristics: 1. It is persistently pleasurable. . . . 2. It is zealously pursued. . . . 3. It is optimistically planned . . . 4. It is begun without thought of profit . . . hence it is an avocation . . . ; but the success may be so great as finally to lead the hobby rider into the professional field. 6. It is individually possessed . . . 7. It makes its possessor free; when working at his profession he is doing what he has to do, but his hobby . . . permits him to do what he wishes."

Beekeeping may easily meet all the characteristics given by Dr. Cutten. If properly managed, it will reimburse the hobbyist for all his expenses. I do not know of any hobby that surpasses beekeeping in present and lasting interest —and I have had more than half a century of it.

E. L. S.

California: August, 1953

Contents

Preface v

CHAPTER ONE: *How to Begin Cottage Beekeeping* 3
   1. Equipment 3
      Information (3)—The Standard Hives, Supers, and Frames (4)—Accessories (6)—Tools (7)—Caring for the Smoker (8)—The Honey Extractor (9)—Clothing (9)—Warning! (11)
   2. Locating Your Apiary 14
   3. Avoiding Stings 16
   4. Summary 19

CHAPTER TWO: *Working among Your Bees* 20
   1. Your First Colony 20
   2. The Four Principal Races of Bees 21
   3. Opening the Hive 23
   4. Finding the Queen 25
   5. Inspecting the Brood Nest 31
   6. Opening without Veil or Smoker 33
   7. Some Random Observations 35

CHAPTER THREE: *The Essentials of Bee Behavior* 39
   1. Success in Beekeeping 39
   2. The Bee Colony 41
   3. Rearing a Queen 46
   4. How to Requeen 53
   5. Requeening without Dequeening 55
   6. Swarming 57
   7. A Few Pointers on Reducing Swarming 58
   8. Laying Workers and Drone-laying Queens 60
   9. Bee Diseases and Their Treatment 61
   10. Other Enemies 65

CHAPTER FOUR: *The Fundamental Principles and Practices of Beekeeping* 68
1. The Three General Principles 68
2. The Eight Essentials of Practice 69
3. The Clear Brood Nest 72
4. Maintaining the Colony in Strength 74
5. Making Nuclei and Dividing 75
6. Uniting Nuclei and Weak Colonies 76
7. Hiving a Swarm 78
8. Preparing Hives for the Honey Flow 81

CHAPTER FIVE: *Harvesting a Good Crop of Honey* 83
1. Getting a Good Crop 83
2. Taking Off Honey 88
3. Caring for Your Comb Honey 90
4. Strained Honey 91
5. Extracted Honey 93
6. Care of the Extracting Room 96
7. Varying Characteristics of Honey 96
8. Rendering Wax 98

CHAPTER SIX: *Wintering Your Bees* 102
1. Man Is the Bees' Worst Enemy 102
2. Are Bees Cold-blooded? 103
3. Preparing Your Hive for the Winter 104
4. Upper Entrances 106
5. Winter Temperature of the Hive 108
6. Feeding 109
7. "Bee Weather" 111
8. The "Warm Way Entrance" 112
9. Supplemental Heat 114
10. Equipment for Warming the Hive 115
11. Setting the Thermostat 118
12. What Supplemental Heat Does for You and the Bees 119

CHAPTER SEVEN: *The Next Spring and Thereafter* 121
    1. Package Bees 121
    2. Your Wintered-over Colony 125
    3. Spring Build-up 127
    4. Conversation Pieces 128
       Beekeepers' Associations (128)—Your County Inspector (128)—Use of Chemicals (128)—The Number of Bees per Colony (129)—A Late Swarm (129)—Syrup Feeding (130)—Good Combs Essential (130)—How Many Bees Can One Man Keep? (130)—How Much Honey Can a Colony Make? (131)—The Twin-Colony Hive (131)—Time of Hiving Package Bees (132)—To Have Nice Wax (132)—Do Bees Damage Fruit? (132)—Pollination (133)—The Location Is Very Important (133)—The Weather Is Also Important (134)—Royal Jelly (135)—Some Facts about Honey (135)—Why Does Honey Ferment? (136)—The Observation Hive (136)—The Dancing of the Bees (138)—Be Openminded to New Things (139)
    5. What Next? 139

APPENDIX 141
    Bee Journals 142
    List of Reference Books 142
    Other Works by the Author 143

INDEX 145

Amateur Beekeeping

CHAPTER ONE

How to Begin Cottage Beekeeping

1. *Equipment*

THE beekeeper's equipment is partly mental and partly physical. Correct information, well absorbed, to which practice in applying it is added, is the essential mental equipment. Then, apart from the bees themselves, come standard hives, with accessories, the right tools for efficient work, and adequate clothing. It is possible to hive bees in any old box, costing you nothing, and learn about working with them afterwards. However, you will not get as good results, and you do not want merely to begin keeping bees; you want to know the best way to begin and how to work the job through to a successful culmination.

INFORMATION. Correct instruction is what the potential expert must have first of all. It is a primary factor in success with bees and will help you to avoid making the many mistakes which you would surely perpetrate without right guidance. For such guidance you have this book; but I would advise you also to subscribe at once to one or more of the representative bee journals. Many of the single issues are worth more than the cost of a year's subscription. There is usually a beginner's department where questions are answered; and if something comes up not covered in this book (the unexpected does happen!), you may find help there. A student or true hobbyist will read anything on the subject of his hobby that he can find. Those who keep bees for utilitarian purposes only may wish to save money by not subscribing to magazines or purchasing textbooks; but in the end this does not pay, either in money or in enjoyment. So—get all the essential information you can, first of all.

THE STANDARD HIVES, SUPERS, AND FRAMES. After reading this book and subscribing to a good bee journal, you are ready to purchase a full colony of bees in the hive of your choice. I recommend either a two-story Langstroth hive with an extra super—which is a third story, usually shallow; or a modified Dadant hive with two Dadant supers. Whatever kind of hive you use, you should have a queen excluder with each hive, a bee escape, and some extra frames with comb foundation to meet the requirements of your anticipated honey flow. These accessories will be discussed on pages 85 to 91.

A good hive must be a good home for the bees; must be so built as to take advantage of and guide the instincts of the colony in producing the best possible crop of honey, and must be convenient and easy for the beekeeper to handle. An old box or the old-fashioned skep would furnish a good home for the bees and even enable them to produce as good a supply of honey as any other home; but they are not as convenient and easy to manipulate as are the modern standard hives, with the movable frames. These were invented in 1851 by the Reverend Lorenzo Langstroth, who, compelled to resign his pastorate because

EQUIPMENT | 5

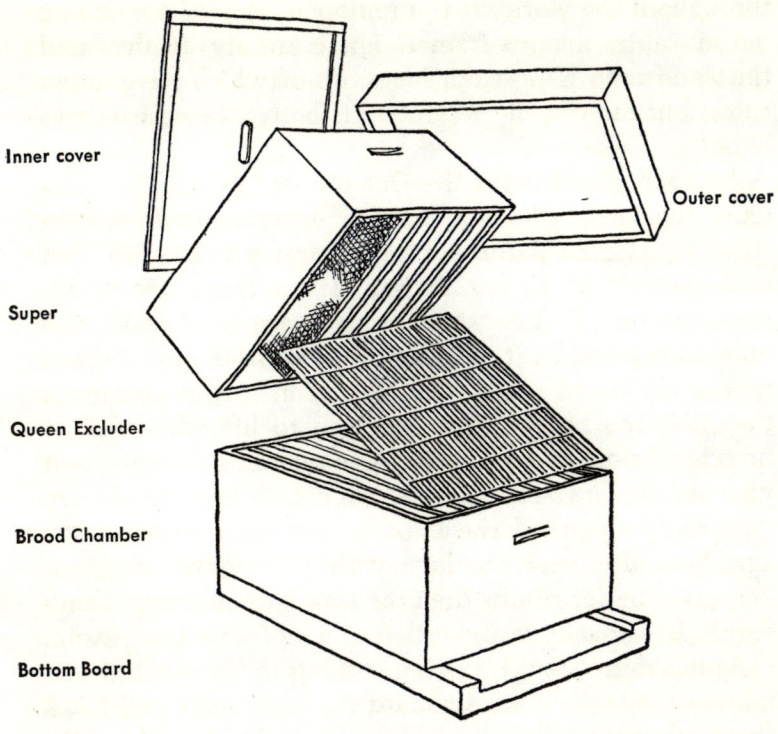

Figure 1
The Hive and Its Parts

of failing health, had to seek other means of maintaining his family. He chose beekeeping. He had a well-trained, scientific mind, and it was not long before he introduced several improvements in hive construction, notably the hanging, movable frames, with provision for proper "bee space" all around. The frames make possible the inspection of every comb and its removal or rearrangement in the hive. This has proved to be the most significant step ever taken in advancing the science of beekeeping.

The Langstroth hive, or some modification of it which may be given another name, has now become standard throughout the world. It is a top-opening hive, containing ten movable, shallow frames. There are also twelve- and thirteen-frame Langstroth hives, all of which have advocates; but for you the ten-frame is better, being easier to handle.

Another good hive is the Dadant, or the Modified Dadant, the result of some years of experiment. It is constructed after the pattern of the ten-frame Langstroth, but is deeper by 2⅛ inches. It has eleven frames, with 1½-inch spacing. If Langstroth supers are used with this hive, as is possible, it is necessary to close the extra space with a couple of cleats. One objection to the Dadant is that it is too heavy for one person to lift when full of brood and honey. If your hive is to remain stationary, and you use shallow, comb-honey supers, it may be a very good one for you to have. Experiments indicate that, averaged year after year, this hive, with its larger brood chamber, gives better results than the standard 10-frame Langstroth. Its capacity is about that of a 12-frame Langstroth.

At this date (1953), the latest thing in hives is the aluminum one. It has the standard measurements and takes standard comb frames and supers. It is light and easy to handle, besides being warmer in winter and cooler in summer. It is somewhat more expensive in the beginning than the wooden hives, but upkeep is less as it never requires painting.

ACCESSORIES. With your hives, you will want frames, of course, and comb foundation. For each hive body you need 2½ pounds of medium brood foundation and one pound of No. 28 tinned wire. You will also require a spur wire imbedder. These are described in your catalogue, and when you buy hives and frames "in the flat," full directions for assembling will come with them. The comb foundation saves the bees a lot of work and time, and starts them off in building even and parallel combs. It is simply an artificially constructed midrib of honeycomb, made of pure beeswax but with heavier walls than usually made by the bees, so as to enable them to employ the surplus in drawing out the cells for the reception of honey or brood. It solves many of the beekeeper's problems, as it encourages the bees to build their combs straight, and to construct cells that are all worker cells. There will be far fewer drones in a hive where comb foundation is used. Moreover, the bees will build their combs on a straight line under the top bars of the frames. Sometimes just a narrow strip of foundation is sufficient to start the bees off straight; but using a full sheet insures the building of worker cells only.

A queen excluder is essential in keeping brood comb away from combs containing honey, and also as an aid in preventing swarming. More will be said about this in pages 58 to 61.

TOOLS. Before you begin to work with your bees, you should provide yourself with a hive tool and smoker. The hive tool has been described as the most useful tool in the world, and it is indeed an extraordinarily efficient, although very simple, little gadget. Of course, one can use some form of knife or a screwdriver, but the standard hive tool is the very best for your purposes, and far superior to any substitute. It is made of high-grade steel, and one end is flat; the other end is bent or hooked and has a V-shaped hole for pulling tacks or nails. The hooked end is used for scraping wax and propolis from frames and bottom boards. Propolis is the "bee glue" which the bees use to weatherproof their hives and to steady the frames of comb and honey. Either the straight or the hooked end of the hive tool may be used for separating the frames. This is done by prying them gently over with a twist which forces them apart with a strong leverage which, nevertheless, does not jar the bees. The straight edge is used for prying up the cover of the hive, separating the super, and other purposes. You will find many more uses for your hive tool as you become expert with your bees.

As important as the hive tool is the smoker. Its purpose is to keep the bees quiet while you work with them. Some instinct leads them to stuff themselves with honey when fire threatens—and, to the bee colony, where there is smoke there is fire. A honey-gorged bee is disinclined to sting.

Printed directions come with all smokers; but in the event that you have a second-hand one without directions, I give a few suggestions as to fuel. In a pasture, dried "cow chips" will do, and peat when it is obtainable. Dried bark, or well-decayed wood, pencil size, will serve also. Greasy waste makes excellent fuel, and also lathe or planer shavings. Dry corrugated cardboard, rolled, is good, and old rags (unless of wool, which is unsuitable). I have come to use mostly bits of old burlap sacking, lightly rolled and just long enough to fit into the smoker. One end of the roll may be dipped very lightly in something inflammable; or, if you want to make up a supply ahead, you may prepare a number of rolls, tying at intervals and dipping at one end in a solution of saltpeter. This will dry and is easily lighted with a match.

Whatever fuel you use, select one that burns slowly. Imperfect combustion is what makes smoke. So do not work your bellows too vigorously: you want smoke and not fire. Too hot a fire will burn the tin off the firebox, and as a result the iron will rust more easily. If this comes about, your smoker will be unlikely to last more than one or two seasons. If well cared for, it will last a long time.

CARING FOR THE SMOKER. Never leave your smoker out in the rain, as your firebox will rust and the bellows fabric will tend to harden and crack. It is a good idea to have a small tool house or weatherproof box in which to keep your working equipment—smoker, veils, gloves, hive tool, etc. The shelf for the smoker should be fireproof, as you might put it away while still hot.

Never allow the gum or greasy soot from the combustion to collect at the top of your smoker until the cap will no longer fit down over the firebox. If you use greasy waste, this is likely to occur. A few moments spent in cleaning the smoker with your hive tool before using it is time well spent. If you await till after using it, it might be too hot.

If the smoker's grate becomes clogged, it may be lifted out for cleaning by inserting the point of a knife or file into one of the holes. When it is removed, you will see the anti-spark tube immediately beneath the grate. This is designed to carry the current of air from the opposite hole in the bellows to the firebox. The sparks cannot work backwards—or outwards to the manipulator's clothing—because the end of the tube extends to the center of the grate. This is a valuable protection which was lacking in the earlier designs.

Instruction in the handling and use of the smoker will be presented in chapter 2.

THE HONEY EXTRACTOR. This piece of equipment will also be considered later, in chapter 5. It is something which you, as a small beekeeper, can get along without but which you may, nevertheless, like to have.

CLOTHING. Next comes the question of what to wear while working with bees. Smooth white or khaki clothing is good. Some people like coveralls with zipper (slide) fasteners. Blue denim is all right after it has been worn and washed often enough to be rid of the odor of the dye. Bees will often sting new blue denim furiously. Do not allow a child with a new blue denim suit to wander among the bees. Do not wear your levis turned up half way to your knees. Roll them down low so that tender ankles will not be exposed.

Do not wear bib overalls with straps. If there is one place where angry bees seem to love to go, it is under those straps or braces over the back and shoulders; and who can blame a bee if it stings when caught there? Under the strap the shirt may be wet with honest sweat, and bees seem to dislike some body odors intensely.

If you make the bees cross, as you well may sometimes, they will invade every crevice in your clothing—between buttons (even entering buttonholes!) and at wrist and ankles. Sometimes they will wander into such places by accident and then sting because they feel trapped and do not know how to get out.

Zippers or slide fasteners are particularly good for keeping bees outside, where they belong.

Nowadays few women would work with bees without wearing overalls or slacks; but it was different in the old days, when a woman always wore a dress. When I was a boy, a cousin once asked me to help her with a few hives on her farm. She wanted me to show her how to handle them. I found out afterwards that she had to stay in bed for two days following our operations; the bees had crawled up under her dress and stung her badly. She had taken it for granted that I was being stung as much as she, but that I was heroically saying nothing; while she was determined to stand anything I could stand. As a matter of fact, I had not been stung at all. My long trousers, tucked into high boots, kept the bees out. It was my first experience in helping a woman beekeeper, and it never occurred to me that her veil and gloves were not all the protection she required.

When I know my bees and find that they are gentle, I usually work without gloves, my sleeves rolled up tight and close. This is much better than wearing a shirt with loose wristbands and plackets, which are most inviting places for the bees to investigate. They love to crawl into dark holes. Also they seem to see black clothing better than white. A person in a white shirt often escapes stings, while his partner, wearing a black or dark-colored one, may be stung badly. Cross bees will sting the black spots on a spotted dog and not touch the white ones.

If the bees are quiet and gentle and the weather is hot, I even like to work in nothing but shorts and sneakers, with a straw hat and a veil. I throw the veil up over the brim so that I can breathe more easily and wipe the perspiration away. (There is now made an elastic sweat absorber which can be put around the head to rest on the brow, preventing the perspiration from trickling down into the eyes.)

WARNING! Do not wear an old felt hat, soaked with sweat, which has been stung so often that it is full of the scent of bee venom. This will arouse the anger of even the gentlest bees.

Some authorities advise beginners to pull their socks up over their trousers legs or to tie a string around the legs. I prefer not to do this, for esthetic reasons: it looks ugly, and there is no reason that one should offend the eyes of others unnecessarily. Besides, it is not a good practice if the socks are woolen, as bees may attempt to sting them. When you are taking off honey or doing some other operation that will scatter bees all over the ground, put on a pair of leggings or puttees, and look neat. You may wear high boots if you like, but these are uncomfortably hot in warm weather. Mostly I prefer low canvas oxfords—sneakers—and when necessary I put on canvas leggings. Leather ones are good, but often too hot.

Do not wear gloves except for rough work. They are desirable in handling hives, when fingers might be pinched, or in moving bees at night (something you, as an amateur, are not liable to do), when any bee that touches the skin will sting. It is all right to have a pair of long canvas gloves in reserve for an emergency; but by all means learn to work with bees, especially your first hives, without gloves.

I have recently found some plastic-coated gloves that give promise of being useful on occasion. They are waterproof and sting proof and wear well, but like all gloves they are somewhat clumsy. Canvas or leather gloves are more readily stung than the bare skin, especially after they have been stung several times. They retain the scent of the venom and cause other bees to attack the same spot. Without gloves, you learn to take hold of a frame without pinching a bee. With gloves, you are less likely to realize how many bees you hurt or kill.

If you feel that you must have something to protect your wrists and prevent crawlers from getting up your sleeves, make a pair of long sleeve protectors of canvas or ticking, with an elastic cord at top and bottom, one small enough to fit the wrist closely and the other larger, to fit over the elbow. These will protect you and stay in place; but they are hot and oppressive on a warm day. Like gloves, they are good things to have when you cannot work with sleeves rolled up, but most of the time you won't need them. When you do use them, always see that they are washed before using them again.

With these sleevelets and gloves and a shirt, which I shall describe, and a good veil, you *might* move your hive at night or meet any emergency without getting a single sting.

You will probably want two veils. One should be of silk tulle, which is airy, light, and fragile, to use "for best," or when you are not in a hurry, or when you are giving an exhibition. The other should be of the folding type, made of four screen-wire panels joined with tape hinges, with fabric above to fit around the hat, and a mesh skirt to tie around the neck under the collar of coat or shirt. This veil folds into a flat package.

With either veil, bees will sometimes crawl underneath

Figure 2

Bee-tight Clothing

unless you are wearing the special shirt which I like to have for serious business. The best beeworker's shirt is an ordinary khaki one of the slipover type, with throat opening large enough to pass over head and shoulders—a zipper front will serve as well. Around, outside, and below the collar, if it has one, sew a standing band about 6" wide and approximately 25" long. Sew a seam about half an inch from the edge, and run a stiff wire through this half-inch space. The band may flop over some, but the wire loop will make it stand out all around. Put the shirt on and snap the elastic cord at the bottom of the bee veil, of whatever type, over and under the collar ring; then tuck the shirttails well into your trousers or slacks. This arrangement is beetight, and the veil will stand away from your face and stay away. No bee has a chance of penetrating this armor! The outfit is too hot to wear for any length of time on a warm day, and you will don it only when you must; but it is effective. Even if you hold a dishpan or box under a big swarm and somebody shakes it down all over you, not a bee will get in.

I have worn such a shirt when working with the bad-tempered Cyprian bees and on emergency occasions when a colony was thoroughly angry and stinging everything in sight. (In bad cases of robbing by other bees, for example, a hive will become so aroused that clusters of them will even try to sting a projecting stovepipe.) The shirt is useful also when you are called upon to help a neighbor who has mishandled his bees and made them so angry that they are trying to run everybody off his place. Of course, if you handle your own bees carefully, all will go well.

2. *Locating Your Apiary*

It is very important to use good judgment in placing your hives. If you locate a hive so that it faces your neighbor's driveway, garden, or lawn, he will probably not consider you a very desirable neighbor. Here are some good rules to follow:

1. Unless you are in a hot, dry, tropical setting, have your hives face south if possible, so that they will have the advantage of the morning sun. An eastern exposure

does not usually give them enough sunshine. Never face them northward.

Do not locate your little apiary facing a much-traveled road, or a field where horses are used for plowing or where they are corralled. I speak from bitter experience, as I once had to pay for a pair of fine farm animals that were stung to death by my bees. They were plowing a field adjacent to my 500-colony apiary.

A good hedge, a high fence, or clumps of trees in front of your hives will cause the bees to fly high as they leave on their nectar-seeking flights. Thus they will be high enough to pass over the heads of people or horses in the neighborhood. In regions of strong prevailing winds, hives should be surrounded by a windbreak anyway.

A friend of mine kept a hive of bees on the roof of his house. This was on the outskirts of a small city which, uniquely—having never been faced with complaints—had no ordinance against beekeeping within the city limits. His rooftop hive was surrounded by a four-foot fence, in addition. He had a good crop of honey almost every year for twenty years, and never received any protests from neighbors; in fact, only a few intimate friends knew that he kept bees there. The bees started out on their flights at so high an altitude that they troubled no one.

2. Do not locate in heavy shade. Light shade is all right. So is an unshaded spot, except in the hottest, driest regions. Do not place your hive near a vine or under a tree with low branches which will catch in your veil; this will result in loss of temper on your part as well as on the part of your bees. Avoid trouble, and keep calm.

3. You may use a hive stand and alighting board as sold by the supply houses, or you may make one yourself, using three bricks or a cement block or two. The hive will last longer if kept a little off the damp ground. There is per-

haps no better protection under a hive than a piece of asphalt roofing about 3′ x 3′ square. This will also keep the grass down. *On any support, set your hive level.* This is to insure perpendicular frames and even, parallel combs of honey.

4. In desert regions, or in any very hot place, the hive will be cooler with blocks under it. Do not have the bricks or blocks projecting beyond the edge of the hive to catch the point of a sickle or scythe or to be bumped by the lawn mower. When using a lawn mower near a hive, remember that bees object to being chopped into small pieces—or to having their sisters so chopped. If the ground is not grassed but is kept clean with a hoe, the bees may resent being struck at, even if your intentions are good. They resent a hoe used carelessly near their hive; they also resent the hands using it and even the person to whom the hands belong, particularly if he is wearing dark, fuzzy, woolen gloves.

5. Put your little apiary where the bees will have easy access to water. This is quite important—more so than many beekeepers realize. In the winter you may have to supply the water, but in summer the bees need to be able to get it themselves. They require it for keeping the hive cool, among other things.

3. *Avoiding Stings*

If a bee crawls up a loose trouser leg, you can sometimes shake it down. If it is on the bare skin and is going to sting, the first sensation the victim feels is a sharp prickling as the bee digs in its toes so that it can hold itself down while it stings. If you are sensitive to that little clutching feeling—as you will come to be, with experience—the thing to do is to slap your hand on the spot instantly, and rub and crush until that bee is no longer a bee with a sting. Bees do not sting the smooth skin readily, but if they

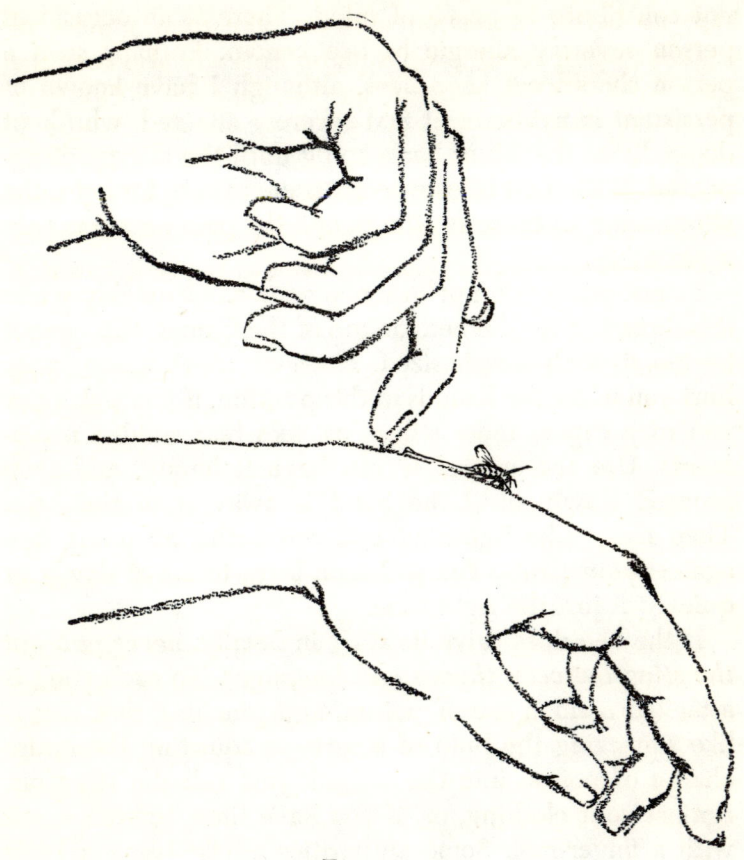

Figure 3

Removing a Sting

do sit down on a bare leg or arm and dig in to sting, I notice the sensation immediately, and quickly kill the bee before it strikes. This soon becomes a habit when working with bees.

While we are on the subject, we might as well have a few more suggestions—adding to your informational equipment. Stinging bees may enliven the occasion, but they do not contribute to peace of mind. There is an occasional person severely allergic to bee venom. Perhaps such a person should not keep bees, although I have known of persistent individuals, at first severely affected, who kept doggedly at the job of beekeeping until the allergy disappeared. With most of us who continue to care for bees, the stings cease to cause swellings, and the pain becomes only momentary.

I must warn you not to move your hand quickly while it is directly over the bees on top of the frames; this cannot be too strongly emphasized. However much a sting may hurt you while the hand is in this position, if you jerk away you may expect more stings, as bees fear sudden movements. Use the smoker if you have it handy, and then proceed slowly until the hand is away from the hive. Then move like lightning and crush the offending bee against your pants. You will soon learn to move slowly or quickly at just the right time.

If the bee does drive its sting in deeply, *never pull out the sting between thumb and forefinger*. To each sting is attached a small sac of poison, and pinching that bag is like squeezing the bulb of a syringe, squirting the entire charge of poison into the wound. Just rub the sting out against your clothing, or, if you have time, scratch it out with a finger-nail. Some authorities advise using a knife blade, but one is usually in too much of a hurry for that. Just a lightninglike movement, and the bee is dead and the sting is out.

4. *Summary*

By now you are beginning to realize that you must have a great deal of new information before you are ready to start beekeeping. You have a good idea of the equipment you will need, your hive and its accessories, and the tools which you will use. You have probably already decided what you will wear as you begin to work with your bees. Also, you know that you must consider your neighbor as well as yourself in the location of your little apiary; and that, in all likelihood, as you work with bees you will soon become accustomed to receiving some stings. The next thing in importance is to learn the steps to take in actual beekeeping practice.

CHAPTER TWO

Working among Your Bees

---

1. *Your First Colony*

*I*T is suggested that you purchase a full colony of bees in the hive of your choice. They should be guaranteed to be free from any of the bee diseases, and to include a young and active queen. By working with your own bees, you will come to know what you may expect from a full colony. You will learn much that you might not learn if you start with a swarm gathered from a nearby tree or with package bees from a supplier. If you purchase a package of bees by mail and have a poor summer, or happen to get an inferior queen, the bees may not build up to full colony strength in a whole season; and as a consequence you would fail entirely in your efforts to secure a crop of honey. Early success is a great stimulus to the amateur.

When to get a package, and what to do with it, will be discussed in the last chapter.

2. The Four Principal Races of Bees

THE KIND OF BEES TO KEEP. There are four principal races of bees. The Italian bee is the most popular, and I suggest that you, as a beginner, obtain these first, as the easiest to manage. Some prefer Caucasians or Carniolans, or even the bad-tempered Cyprians, but on the whole the Italians give the best results. In my opinion, the old-fashioned dark Italians were the best of all—and I had ample experience with them during a seven-year sojourn on the tropical island of Tahiti; but these may be found today only on some such island, although there are still relatively dark and light Italians. They are the most important race commercially, good workers, and generally docile to management. The queen will usually vary a little from the workers in color as well as in size, and so is easier to "spot" than are the queens of some other races. Usually the upper part of her body is yellow, while the tip is black. Sometimes she has alternate black and yellow bands. The Italians are less ready to swarm than any other race, although they *will* swarm when conditions are right.

The Carniolan is a silver-gray bee, the segments of the abdomen being black, banded by a grayish-white, fuzzy ring. Some beekeepers think the Carniolans are gentler than the Italians. They will often rear brood under conditions that would stop the latter. They deposit little propolis, and so their combs are exceptionally white and clean; but they have a serious drawback in that they are excessive swarmers.

The Caucasians, similar in appearance to the Carniolans, are probably the gentlest bees in the world, and seem to be gaining in favor. But they are heavy producers of propolis, staining their combs badly, and so are not suitable if you want nice-looking comb honey. They sting very little, and almost never require any smoke when their hives are opened. They are heavier swarmers than the Italians, but less so than the Carniolans. Both Carniolans and Caucasians are resistant to European foul brood—a bee disease which you will learn about in pages 61 to 65—and they are not inclined to run to the edges of the combs when the hives are opened, as the Italians will unless very gently handled.

The Cyprians resemble the Italians in appearance, but are a dark orange in color, instead of yellow. They usually have three orange-colored bands, with brownish fuzz bands between, all bordered with black. The distinguishing mark of the Cyprians is a yellow shield, the "crescent," at the base of the thorax. The Italians show a trace of this shield, too, but it is much less marked. The abdomen of the Cyprians is a lovely orange shade, while that of the Italians is a muddy, dark brown—except for the queen, who may have an all-yellow abdomen, although this is rare. The Cyprian bees are not obtainable from queen breeders in America, but may occasionally be found among beekeepers who are doing experimental work.

Many fanciers like the Cyprians for their beauty—they are undoubtedly attractive—but they are exceedingly temperamental, being the crossest European bees known (the giant bees of India are doubtless worse). They are so very vicious that few people have the temerity to accept the challenge they offer. Sometimes every bee will leave the hive—no amount of smoke will stop them—to make a fiendish attack on some fancied enemy.

There are other races, but those I have mentioned are the most notable for the American or European beekeeper. If you want the glamorous Cyprians, after this description, never work in your apiary without the most bee-tight armor you can get. And place the hives a long way from family and neighbors!

3. Opening the Hive

Having obtained your hive of bees and located it in a suitable spot, the next step is to learn how to work with it. To acquire a knowledge of what to do and what not to do, read the following pages several times. Do not be so eager to go to work on your new possession that you will try to open the hive before you have the instructions and precautions well in mind.

FIRST BECOME ACQUAINTED WITH YOUR TOOLS—THE HIVE TOOL AND SMOKER. Try the hive tool out as a pry, and as a scraper. Play "jack straws" with it, prying things apart with as little jarring as possible. If there is no suitable strap or pocket in the outfit you are going to wear when working with your bees, see that one is added which will hold the hive tool conveniently at hand. Know just where it is, always.

Then test your smoker. If the smoke seems too hot, a little fresh green grass on top of the fuel will cool it down. The first commandment for the beekeeper when working is: *keep your smoker going*. If you have only one hive, you should have no trouble in doing this; but even as few as a dozen hives worked at one time will keep the operator so busy that only long experience will develop the habit of remembering this behest. Impress it on your mind by frequent repetition.

Also study and memorize the procedure for opening the hive, finding the queen, inspecting the brood nest, or doing whatever you plan to do, before you go to work. If you haven't impressed the important steps upon the memory by frequent repetition, and so are not sure just what to look for, you may be tempted to set down your lighted smoker and look in the book to see what to do next. Then your smoker will burn low or go out, and you may attempt to resume work without enough smoke. As a result, you may find yourself retiring in disorder, ripping off your veil to get the bees out of your hair. Or, if you have on absolutely bee-tight clothing, you will at least make the bees so angry that they will be no fun to work with.

THIS IS THE WAY TO OPEN YOUR HIVE. It is a warm, quiet day in early spring, and you have recently acquired your colony. Avoid working on a windy or stormy day. As yet, your hive has no super, but its activity would indicate to a seasoned beekeeper that brood rearing was going on apace, and that a super would be needed very soon. It is between 10:00 A.M. and 3:00 P.M.—the best time for working with bees; and it is preferably at an hour when you can arrange your position so that the sun will not be in your eyes as you examine the combs. If you have partial shade for your hive, all the better. You are suitably dressed, and you have your hive tool and smoker in hand, the latter going well but not so well as to make a flame. Avoid standing in front of the hive entrance. Stand to one side, and blow a *very little* smoke on the alighting board *but not into the hive.* (After more experience you will not normally use any smoke at the entrance.) Beginners are inclined to give too much smoke; so be conservative.

After this first bit of smoke, insert the flat end of your hive tool under the top cover and remove this carefully, so as not to jar the bees in the least. There may be cracks or openings in the inner cover if your hive is old, and you would puff a little smoke over these first. Then insert your hive tool under the edge of this cover, and lift it very slightly, taking care not to snap the propolis any more than can be helped. Puff a little smoke between hive and cover. Do not blow it down between the combs, but allow just a whiff of cool, white smoke to drift over the tops of the frames. With the first puff you will notice almost at once a difference in the quality of the hum of the bees inside the hive, and when this settles to a quiet, uniform buzz, as it should in a minute or, at most, two minutes (if not, give another light puff), you may open the hive. Experience will soon tell you when it is safe to lift the cover. Do not use enough smoke to make the bees logy, especially when the day is very warm and the colony buzzing with activity. Often on such a day no smoke at all is required, as the bees are so happily busy that they are, as it were, intoxicated with success and will pay little or no attention to you. However, while you are still an amateur, be cautious: use your smoker, but sparingly.

Consider opening your hive successfully your first accomplishment, and rejoice in your developing skill.

4. Finding the Queen

If not the first time, then the second time you open your one-story hive, let your object be to find the queen. If you use too much smoke, you will set the bees to running so wildly over the combs that locating her may become an impossibility. After you have removed the top cover, you may, if you wish, turn it on its side and sit on it, balancing yourself there while you take off the inner cover, and gently separate the frames with your hive tool. This permits quite a flexible arc of movement for your body and arms, and is not too uncomfortable for a brief period.

Your smoker is beside you on the ground, and even if you do not use it again on the hive—as you may not if you find the queen promptly—you will pick it up every minute or two and give the bellows a squeeze to keep it going. Only if you detect a change in the buzzing or behavior of the bees—indicating a rising restlessness—will you find it necessary to give them more smoke.

Let us say that you want to remove the center frame first. A slight twist of the hive tool, using the end which seems most suited to the action, is enough to separate the frames without jarring them. One at a time, beginning at the side nearest you, you gently crowd the frames toward the side of the hive until you have room to secure a firm hold with both hands on the frame you want.

If you attempted to pull one out without first separating them, you would be in danger of rolling bees over one another and crushing them. This would not only result in angry bees, but might even kill the queen.

It is a good plan, after removing the first frame, to crowd the rest together in pairs, two frames quite close together with only the "bee space" between, then a wider space, and so on. The queen is then likely to go between a pair, where it would be a little darker.

Before removing the frame, however, lay aside your hive tool, or put it in its strap or pocket of your coveralls. You may learn to keep it in your hand, held by your little finger, while handling the frames, but on this occasion it is well not to run the risk of hampering your movements. Gently lift the frame out, slowly and carefully, using both hands. Do not allow it to touch the end of the hive. There is only a "bee space," $5/16$ of an inch or a little less, between the end of frame and hive, and there will be bees in that space, pretty well filling it. So draw up the frame carefully, straight up, so as not to crush a bee. If you wear gloves, your fingers will be so insensitive that you will almost certainly injure bees without knowing it, and when a few bees begin to "yell" that they are being hurt, others will rally to their defense, and then you may expect stings.

Act deliberately and without nervousness, lest you unconsciously move jerkily and cause the bees to reciprocate your own fears. After drawing the frame out, first look on the side of the comb in the frame next to the one just removed: it is barely possible that you have smoked the bees too much, and so disturbed them that they have run to the sides, followed by the queen. This quick look at the adjacent frame will tell you what has taken place better than looking at the one in your hands, for moving this may have changed the situation somewhat. If the bees have run to the sides, I suggest that, since you are a novice, you replace the removed frame and take out two or three of those next to the sides, depositing them gently against the side of the hive outside; and then slowly even up the separated frames in the hive in more distinct pairs—even though you may already have paired them slightly. When this is done, replace the cover and wait a few minutes. If you wish, you may examine the frames which you have removed, but you are unlikely to find the queen there. If it seems necessary, you may pass a light puff of smoke above these frames while waiting. After about five minutes try again, this time taking the paired frames, one at a time, and examining only the sides facing each other. You will probably find the queen without difficulty, but if you do not, replace all frames and try again another day.

If you are using the Hoffman, or other self-spacing frames, it is not necessary to separate them around the outside of the hive. You may even take them out in groups of two or more and leave them sticking together as you lean them against the outside of the hive. The bee space obviates any danger of killing the bees between the frames. However, if you are using unspaced frames (without the spacing shoulders), you must carefully space them 1⅜ inches from center to center. This is quite a nuisance, because it means irregularity in thickness of combs; so, if you possibly can, use the Hoffman or similar frames.

Instead of leaning the frames, of whatever kind, against the side as you remove them, you may have a super beside you on the ground, as a carrier in which to place them. If there is danger of robbing from other bees in the neighborhood, it is well to do this anyway and to cover up the frames as you place them in the carrier. Experience will tell you which you prefer to do. Later, you will want the carrier, anyway, when you are taking off honey.

If all has gone well, and the bees have not run to the sides, hold the removed frame with top bar horizontal for easy examination. If it is a comb containing eggs and very young brood—unsealed—the queen may be there. If she is not, turn the frame over. If it is light—not heavy with honey or sealed brood—it may be turned right over, with the bottom bar on top, still horizontal. If it is both heavy and unwired, raise the end in your right hand—your left hand remaining stationary—until the top and bottom bars are perpendicular. Then rotate the frame around, as if it were on a pivot extending up from your left hand, until the opposite side is facing you. After examining the frame and finding the queen there, replace it and close the hive.

If she is not there, lean the frame against the side of the hive or place it in the carrier box, and remove another one, following the same procedure. This may be leaned against the hive separately or aligned with another Hoffman frame, or it may be returned to its place, leaving space for easy removal of the next frame. Proceed in the same way until you find the queen, or conclude that both you and the bees have had enough for this first attempt. *Always replace your frames in their original order*—until you learn a reason for not doing so.

Have you kept your smoker going?

As this is your first lesson in practice, you may congratulate yourself if you have located the queen; if you have not, you will by all means succeed eventually. However, in practical beekeeping, it is seldom necessary to find the queen. You will know she is there if brood is present in all stages of development (study Figure 4 carefully so that you will know how to judge the age of the brood) and if the population of the colony is well maintained. It is not often required that there be further evidence of her presence. So, if you see the results of her work as you care for your hive, do not be concerned if you fail to find her on any particular occasion. Later we shall discuss circumstances when you may desire to replace her or to find out what is wrong.

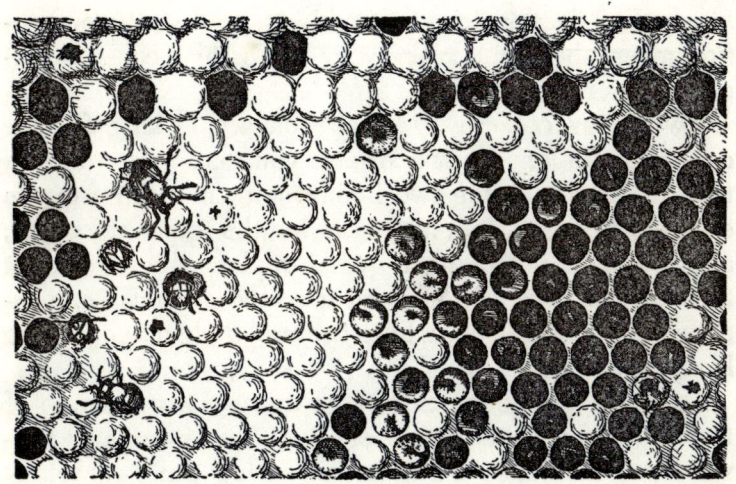

Figure 4

Brood Nest
Large Cells at Top Are Drone Brood,
Eggs and Larvae at Right, Sealed Pupae and
Emerging Adults at Left

When you make this first examination, the queen will be busy; so make your search not on a comb frame containing little or no brood, or containing a lot of brood all sealed over, but on one containing eggs and young brood. You may usually remove the comparatively empty frames and put them aside with only a cursory examination.

Since you may have only one hive when taking your first

lesson in management, and since the bees are busy, there is little likelihood that another colony will start robbing your open hive—a thing which may happen when there are several colonies, and when nectar is scarce after a hot, dry spell. So it is quite safe for you to set the combs outside the hive while examining another one. However, if, when you have built up your apiary, you must examine your hives when there is danger of robbing, observe the precaution given above and place your combs in a covered carrier as you remove them.

TWO-STORY COLONIES. When you are looking for the queen in a colony of two stories, you may use your hive tool first to loosen the two bodies, and then slip a queen excluder between. You may examine the upper story first, and if you do not find the queen, replace the covers, lift the super off carefully, and put it on the ground. Then examine the lower body, where the brood chamber is more likely to be.

Or, since the queen cannot go up or down because of the excluder, and ten to one she is in the lower story anyway, you may take a chance on saving time, especially early in the season, by lifting off the upper body before examining it, setting it apart quickly, and covering it, while you work the lower story first. Later in the spring, or in the summer, you may find the queen more readily in the super, unless you have had a queen excluder on to confine her below.

There are almost failure-proof ways of finding the queen in a hurry in a many-storied colony. One is to smoke the entrance heavily and drum on the sides of the hive. This makes the queen and a lot of bees run to the top, or to the queen excluder if one is used, where she can easily be spotted. This is a good procedure for the commercial beekeeper, when necessary, but it hardly seems sporting for the hobbyist. Besides, it is bad practice when the bees are busy on a warm day, as it confuses them and practically puts a stop to their operations, particularly their flights for nectar and pollen, for the time being.

From what you read here, and from experience, you will come to "sense" the times when more smoke is indicated. When you see, between the frames, the black, shiny heads of a lot of bees, with wings half raised, poised for taking off, that is the time to use just enough smoke to make them put their heads down and go to fill their bellies with honey to carry with them just in case this smoke turns out to have fire behind it. (This must often have happened to their wild ancestors in the forests.) You might smoke them even a little earlier, when these watchful, suspicious clans of guards are gathering. The beekeeper must know the course their age-old instinct leads them to follow, and take advantage of it. They seem to know that a fire may prove dangerous, but with each guard bee full of honey, they swarm out, go to a new place, and start a new home. Full of honey, they cannot well bend their bodies to sting, nor do they seem inclined to do so. They are bent only on preservation of the colony, and as they prefer to keep it at its old stand if possible, they will not swarm out until just about the last minute to escape fire. So never let your smoke turn to flame.

5. *Inspecting the Brood Nest*

To have control of the queen, and therefore of the activities of the colony, the honey getter must know accurately at all times the condition of his colonies and the direction the bees' activities are taking. Inasmuch as the life cycle of the bee involves many changes within a short period, the operator must know at intervals of 8 to 10 days (whether or not he works every colony that frequently) what changes and developments are taking place in each colony under his care. Examining the brood nest of many colonies as often as this means a great deal of work for the commercial beekeeper; but for you, with only one or two hives, it will be a comparatively easy matter. Moreover, with experience you can often judge colony condition by the behavior of the bees at the entrance. It is because of this that the expert honey producer develops a plan of management which requires a minimum of brood nest examinations.

The brood nest is that part of the brood chamber in which are the eggs, young larvae, the younger workers (nurse bees), and the queen.

Having opened the hive according to directions already given, go through the frames one by one and see how the brood nest is arranged. It should be compact, shaped roughly like a football, with the least amount of brood in the outer combs. It may be in from 2 to 10 combs, according to the season, with the central frames almost full of brood. At the peak of the breeding season, the brood nest may extend into the upper story, if the colony is in a two-story hive. There may be exceptions, however. In the brood nest there will be eggs and larvae of all ages. Some patches will be capped over, others not yet ready to be capped. From still other cells, young bees will be emerging, cutting their way through the brown cappings. Some combs will contain yellow, red, or brown pollen and cells of nectar and unripe honey, as well as cells of sealed honey. All these will be arranged in more or less concentric circles around the center of the brood nest.

Let me repeat this: after you have examined a frame, return it exactly to where it was before you took it out. This is particularly necessary in the cool weather of spring, when the small brood nest is expanding rapidly and you have no arrangement for supplying your bees with supplementary heat to keep the whole hive warm enough for brood to thrive anywhere in it. At times, putting a comb in the wrong place might cause swarming.

6. *Opening without Veil or Smoker*

When you want to open a hive without veil or smoker, a different procedure is involved. For one thing, try it only on a warm, pleasant day, when the bees are happily busy. Then use a little "psychology," which works as well for bees as for humans. Put the edge of your hive tool under the hive lid *at the side opposite you,* so that when the first bee looks out of the crack, it will see no unusual or disturbing sight. It will not see you at all. The first crack merely lets in light, and is not wide enough to permit a single bee to fly out. You move the cover so little and so slowly that there is no snapping of propolis to alarm the bees. If you dislike stings as much as I do, you will be very careful. Do not tear off the lid with a jerk, admitting a sudden rush of air—this is dangerous, even with a smoker! Slow and easy is the word. Remember not to let a single bee escape and raise a row. The bees then see that nothing is happening to them; that the sliver of light is doing them no harm.

When they have adjusted themselves to light and air from this new place, lift the lid a little higher, *if they have remained quiet—not otherwise!* The secret is to accustom the bees to these new experiences gradually. If they show restlessness, stop and wait a little. Remember that you have no smoke to defend you and that your safety depends on persuading the bees that nothing that is happening will harm them. If they remain quiet, not flying out except about their regular business, you may lift the lid a little higher. If they still remain quiet, as they probably will if you move slowly and do not call their attention to yourself, you may remove the tops entirely and expose all the frames.

They will still be watchfully waiting to see whether anything is going to harm them, and the next step is to move your right hand (unless you are left-handed) with the hive tool in it, very slowly over the hive, watching to see that no bees fly up. If one does, it may settle on your hand without stinging, if you hold very still and do not flinch; but woe to you if you are nervous and jerk your hand away, or even twitch a muscle! If a dozen bees fly up, you would best close the hive and retreat as quickly as possible. Something has gone wrong, and more will follow the first ones.

If all has gone well, then pry the frames apart a little at some promising place, usually the second frame from the side of the hive nearest you, so that you do not have to reach over the hive at first. Take the second frame rather than the first because the comb next the hive wall is often fastened firmly with brace combs that would have to be broken. You can usually pry this first frame a little closer to the wall without disturbing the bees badly. The second frame will ordinarily have a good bee space on each side of it. However, use what good judgment you have: more will come from experience.

Pry the frames apart, giving a little more space at each side and moving the hands carefully and slowly. If even one bee flies up, "freeze" the hand motionless so that the bee will either fly harmlessly away or, if it does alight, will have no reason for stinging but will sit quietly, not even "digging in."

This requires self-control, but once this is gained, you will have much self-confidence and nerve. After you have proved it, it is all done so easily and instinctively that your precautions are not observable by the onlooker.

It is worth while to devote much time to learning these

small points, because on them depends your ability to handle bees with ease and a minimum of stings. With one frame out of the hive and exposed to the light, don't be in too much of a hurry to go farther. Let the bees become used to the empty space you have made, particularly if it is a two-story hive; because the bees in the lower story must also become accustomed to the light and movement. Then you may set the removed frame down against the hive—not in the sun, not crushing any bees—and take out another frame or do anything you want to do as if you had smoke, always being careful to let nothing slip and not to bump or jar the hive. Bumping and jarring are permissible only if you are using the smoker at the same time.

After some experience, you will be surprised at what you dare to do without smoke and at how little smoke you really need if you use a small amount at just the right time.

7. Some Random Observations

A successful commercial operator will work out for himself, after he has learned through experience the habits of the bees, a method of management which gives him adequate control of colony activity with little labor. However, this is not for the beginner—who needs to know many things, who asks himself why this or that occurs, and who must go to the bees for the answers. He can secure these answers best by working with and studying his few colonies, and if he wishes detailed information on any point, he has the standard authorities to consult.

When there is honey coming in, or if there is a tendency for robbers to intrude, the guard bees will be very watchful, and it is well at such times to use a little smoke at the entrance. Then a puff into the crack at the top just opened will prevent any bees from flying out when the inner cover is gradually raised and lifted off. There are even times when it can be ripped off at one jerk, with only a puff of smoke, and the edge rapped on the ground in front of the hive so as to jar off all the bees in a heap. It looks like rough business, but the bees stand for it when the time is ripe and you have opened the hive properly. Indeed, a hundred or two frightened bees, tumbled suddenly in front of a hive and running into it, so disorganize the guards and the whole hive that little trouble is to be feared from that colony while you are working on it.

However, if you blunder at any time, never bank on the bees doing anything but attacking you. *Always keep that smoker ready.* A good place for it is between the knees as you stand. This means that you should have a good-sized smoker with an insulating shield for the hot firebox. Otherwise, trousers and knees may be burned. In the old days, before smokers were made with shields, the badge of the beekeeper was a burned area on his overalls at the inside of the knees. After some experience you may move around a few steps with the smoker held there, never thinking about it.

Also, you may keep your hive tool in your working hand, the little finger clasped over and holding it, leaving the other fingers and the thumb free for handling frames.

Do not lay the hive tool down on some hive top or drop it on the ground. Keep it in your hand or in its own pocket or strap, ready for use.

Now that you have learned how to open a hive and handle the combs, I advise you to go through the procedure frequently at first, to familiarize yourself with the colony and its varying behavior under different circumstances. Work at least once a week—every day if you like, or even several times the same day, early in the morning and late in the evening. I advise a would-be commercial beekeeper, who expects to work with hundreds of colonies, to work when it is hot and sunny and when it rains; because sometimes queen rearing or other jobs will make it necessary to work in bad weather. However, you will not have to do this as a routine practice, for with only a colony or two the time factor is not very important.

If you open a hive after dark, you may expect to receive some stings unless you have on bee-tight clothing. When you do try it, keep lantern or flashlight some distance from the hive, and preferably do not let it shine on your body. Bees will fly toward the light, or a light spot, and if they chance to alight on bare hands or face, they usually sting. Then is a good time to wear gloves, and the special bee shirt described on page 13.

Handling a hive so frequently, especially by an amateur, does not produce good results in honey getting; but honey is not your chief concern just now, when you are going to school to the bees. If you do get some honey, well and good; but first learn how to handle bees by working them repeatedly under all conditions.

If you have an opportunity to work for some beekeeper who will give you instruction, it will be a good experience, even though you work hard and get stung often because of the careless and hurried handling common in commercial apiaries. You may, as a beginner, get more than your share of stings, but what does that matter? You will learn much just by repetition; and practice makes perfect, especially if you know what the end result is to be. Wrong practice often leads to bad habits which are hard to break, unless you know what is wrong and why. If your boss tells you to do something which you know to be quite wrong, do not be dismayed. Just do as he says, and see what the results are. Maybe they will be good under the special conditions of time and place, and you will have learned something; or, if they are really wrong, you will find out the reason. Just now you need experience, and you may well gain it by working for somebody else for a while.

Later, when you return to your own hives, you will disturb your bees as little as possible if you desire honey and profit.

CHAPTER THREE

The Essentials of Bee Behavior

1. Success in Beekeeping

*N*ow that you have your own bees and have successfully opened a hive—whether to find the queen or to inspect the brood nest—and have had experience in handling hive tool and smoker, before taking further steps you should acquire some basic information about how bees normally act. For, unless you understand *why* this or that procedure is recommended, or another discounted, you will find it hard to remember what you are told to do. Moreover, some knowledge of what motivates your bees and of the normal functioning of their instincts will give you a foundation for taking advantage of their ways of doing things—to your profit.

The hobbyist wants to know everything possible about his subject and even tries to discover new things. However, for practical purposes, the beekeeper does not need a complete knowledge of the anatomy and physiology of the bee; so I shall not attempt to give in this book more than those facts which every beekeeper must have in mind in order to correlate his work with the activity of his colonies. For success, he must plan his work on the basis of principles rather than on random, rule-of-thumb methods.

As stated in the Introduction, any problem involves three factors: end, or purpose; cause, or plan; and effect, or the results achieved. A good marketable crop of honey is the effect which a commercial beekeeper wishes to secure. The hobbyist aims to make his knowledge useful, also; but he is more concerned with becoming an expert and less with making a living.

If a man has five colonies of bees near a clover field and has secured a good average crop of honey from each hive, he could conclude that the *cause* was the proximity of his bees to the clover. That would not be the whole truth, however, as it might happen that one colony would store twice the average of the lot, while another would produce scarcely any honey at all—for many other factors are involved.

Realizing this, he might draw another hasty conclusion —that the colony which produced so much honey was of an especially good stock—and requeen his other colonies from the queen of the most productive hive. Again this might be a mistake, because he does not know the quality of the drone which mated with the queen. This is an error which has been made over and over again by beekeepers.

So, without scientific experiment, we cannot choose one factor at random as the cause of success and say: "This is it!" The queen of the very productive colony may have been from an excellent strain herself but might have mated with "any old scrub." Her worker daughters, like herself, might then be of fine quality, but her queen daughters might be the poorest kind of breeding queens; while another rather indifferent queen, mated to a drone of fine producing stock, might produce queen progeny much superior to herself.

In general, success in beekeeping is the result of a wide knowledge of bees and of beekeeping practices and the ability to use that knowledge intelligently.

2. *The Bee Colony*

The words "colony," "swarm," "hive of bees," or "stand" are all used interchangeably to describe practically the same thing. A colony may consist of 50,000 to 100,000 bees, more or less. There is normally one queen among them, and a few dozen drones in summer but none in winter. All other members of the colony are worker bees, incomplete females designed for the production of honey, not progeny—although without them the queen could not raise offspring, either; and thus, in a sense, the whole colony is the mother of the hive of bees.

The queen is the only fully developed female and is the mother of all the other bees of the colony—which is no better than its queen. If she lays 1500 eggs a day, she is a good queen, but a better one may lay more. On the number of eggs the queen lays and on the length of life of the worker bees, depends much of the productivity of the colony. The queen has a longer body than the worker has, although structurally they are much alike. Some students state that the queen, too, is not a fully developed female, because she cannot care for her young but must depend on the worker bees for help. Her reproductive organs, however, unlike those of the workers, are fully developed. She merely shows no maternal instinct toward her offspring. This instinct is much more apparent in the workers, who feed the young, care for them, and defend them at the cost of their own lives if necessary. The queen is really just the matrix of the colony—an egg-laying machine.

A good queen lays eggs in regular, football-shaped cycles. Starting at the center of the comb, she deposits her eggs in spiral upon spiral until the area of brood covers almost the whole frame of comb, with just a border of honey and beebread around the edges. Beebread, by the way, is just pollen mixed with saliva from the bees, or a little honey, and is stored through the entire breeding season, especially in the spring and fall. It is consumed by the nurse bees in feeding the young brood.

The brood develops and emerges in the same crescent-shaped order in which the eggs were laid. It requires altogether about twenty-one days from egg to bee: three in the egg, six in the larval stage, and twelve as sealed brood. This may vary slightly with the weather.

The worker bees do all the work of the hives except laying eggs, although on occasion they may even do that, adding much thereby to the beekeeper's confusion. As they are imperfectly developed females, and have never been mated, their eggs develop into drone bees. This is called *parthenogenesis,* or virgin birth, and occurs also in certain worms and crustaceans and in some other insects. It is said that in rare cases another female is produced. But I have never come across such a case nor met another beekeeper who has!

The young worker bees spend the first life period inside the hive. This covers about twenty days, but the young bees are not idle during this time. They prepare cells for the queen to lay in, watch for the eggs to hatch on the third day, secrete the food for the newly hatched larvae from glands in their bodies, and care tenderly for the fast-growing larvae until they reach a point of development on about the ninth day, when no more food need be given to them. The young workers then cap the cells and go about other duties while the larvae continue their development until they are ready to emerge as full-grown bees on about the twenty-first day, depending particularly on the temperature.*

After about three days spent in cleaning out cells for the queen, and nine or ten in care of the larvae, the young bees turn to other work in the hive. This consists largely in secreting and working wax. Before they become field workers, they seem to secrete wax readily from the raw nectar with which they fill themselves, obtaining it from the field bees as they come in. With this wax they build combs of two kinds of cells, the larger ones being drone cells, in which drones are reared. Drone comb may also be used for storing honey. It requires less wax and can be built more quickly than the worker comb, which is likewise used for honey storage.

The young bees also help in the evaporation of the nectar, which usually comes in very thin. The hive bees take it into the "honey stomach," regurgitate it, and manipulate it in their mouth parts—a process comparable to "chewing the cud," except that, when prepared, the honey is deposited in the cells. The fanning of the wings of the hive bees helps in further condensation, and it is finally sealed over. The enzymes in the nectar and in the bodies of the bees also have something to do in the process of changing nectar into honey, for they invert the sugar, which may be largely sucrose in the nectar, into dextrose and levulose.

Sometimes the young bees may be seen having "orientation" flights around the hive, when they appear to be playing with each other during the sunny part of the day. The flights usually commence during the latter days of the nursing period, which sometimes extends beyond thirteen days if there are not enough young bees. Often their play

---

* Giving the colony "bee weather" under any external condition is a recent development. The method is described in *Scientific Beekeeping* by Sechrist and MacFarland.

becomes so noisy during a season of heavy breeding that the novice may think a swarm is gathering. When the young bees have made enough such flights to find their way back into the hive, they start carrying out debris—cappings removed from brood cells, dead bees, etc. After this there is a brief period of guard duty, and then they take to the field to gather and bring into the hive pollen and nectar from the flowers.

The nectar is stored in the cells to be changed into honey, as described above, and the pollen is also stored to be made into food for the larvae. The bee carries its load of pollen in two small receptacles, called "pollen baskets," on the outer part of the hind legs. After being mixed with saliva or honey it is stored in the worker cells of the comb and gradually takes on the color of the flower from which it came. Sometimes each color seems stored in separate cells, but a cross section will show that the bees make no effort to separate and store loads by colors.

While the worker bee may live less than six weeks in the busy summer season, bees which emerge in the fall live until the following spring, if they are fortunate. When it is not too cold, and if they have enough food of good quality, they do not die until they have more than replaced themselves with young bees to carry on the springtime routine of the colony—which is what the beekeeper wants them to do. If they do not accomplish this, the colony dwindles and may die out entirely. This is called "spring dwindling" and is by all means to be avoided.

The worker's life is shortened by the wearing out of its wings. You can easily spot hundreds of ragged-winged workers dragging their weary bodies into the hives at the end of a warm, busy day and readily distinguished from the younger bees with their perfect wings. The young bees are also downy, whereas the field bees soon wear off their early fuzz.

Besides the pollen and nectar, the bees also gather the gummy exudations of trees and plants, which they convert into propolis, mentioned on page 7. They use this freely to seal up the cracks in the hive and to cement the frames together. It discolors the hands of the operator and is sometimes quite a nuisance; but it is easily removed with gasoline or turpentine.

Figure 5
Lower—Bee with Pollen Baskets Full
Upper Left—Old Bee, Upper Right—Young Bee

The drones are the male bees and have no function but to mate with the queen—and only one of them can do that! They are somewhat larger and stockier than the other bees. They may be reared in the spring and early summer months, and will be tolerated in the hive only as long as the colony is prosperous. When the honey flow comes to an end, the rearing of drones ceases, too, and the young drone brood will be cast out at the entrance, while older drones will be pushed out to starve, as they cannot feed themselves. If the colony is queenless, or the queen is obviously failing, drones may be tolerated in the hive longer than otherwise.

One of the remarkable things about drones is that they are hatched from unimpregnated eggs. A black queen will produce a black drone, regardless of the color of her mate. The drones that are sometimes hatched from worker eggs are smaller than usual, and it is not certain whether they are capable of fertilizing a queen. There seems to be evidence on both sides of the question.

The behavior of the workers toward the drones may be an indication of the state of the honey flow. If a worker is observed buzzing on the back of a drone that is trying to get away from the hive, it usually indicates that the honey crop is lessening.

Drones, by the way, have no stings. So do not fear their loud buzzing.

3. Rearing a Queen

The drones have no function whatever in the hive, so far as can be determined. Perhaps they support the colony morale by completing it as a reproductive unit. For us, their function lies outside the hive, not in it. They often congregate from large apiaries by the thousands and may be heard high up in the air, out of sight. Then, when a virgin queen flies out on her mating flight, she is attracted to this multitude. One of them mates with her and dies immediately, while the queen returns to the hive and seldom emerges again unless with a swarm from her colony—and she is unlikely to do this the first year of her life.

This matter of uncontrolled and indiscriminate mating is what has made it difficult for the beekeeper to breed a selected strain of bees, or even to keep the races pure. Queen breeders find it difficult to furnish queens of uniformly fine quality year after year because of this uncertainty in the mating of their breeding queens. However, now that artificial insemination is possible, the future may see an improvement in our strains of bees, although research is still in an experimental stage.

As an amateur beekeeper, you will not really need to do any queen rearing, as you can order a queen when you need one from some good commercial producer. However, as a hobbyist, you will want to raise a queen or two of your own occasionally. That will be a good opportunity to test your success against the results given by a purchased queen. All good beekeepers who raise their own do so from their best producers, in spite of the risk. It is not local conditions alone that make a good producing colony and, other conditions being equal (management, for instance), the only other factor must be the queen. Records have been kept of queens from the same stock in widely separated localities, and one stock has been found to produce poor results no matter where located or how expert the management, whereas colonies from queens of another stock would uniformly produce more honey than other hives in the same yard.

The qualities to look for are gentleness and productivity, and on the average you will do well enough if you breed from a queen whose progeny manifest these characteristics.

However, whatever you may do later by way of experimentation in queen rearing, in the beginning you may merely allow nature to take its course and permit your colony to build its own queen cells. If the colony becomes overpopulated, or if the queen for any reason begins to slow up in her laying, queen cells will be built. These are longer and larger than the cells of the other larvae, and the development of a queen seems to depend entirely upon the size of the cell and the difference in the amount of food offered to the little larva. You can promote the development of a queen yourself by putting a frame of eggs in a colony that has no queen. When they hatch, some of the cells will be given a larger amount of the milky food (made from pollen) which is furnished to the larvae than will others. These well-fed cells will then begin to be enlarged, drawing from the adjacent cells, which will become smaller. The queen cells will become peanut-shaped and take up as much room as three or four ordinary cells. After sealing, the young bees put an excessive amount of wax on the queen cell and indent it so that it looks quite like a peanut—or a cross between a peanut and a thimble!

If you, yourself, do not requeen, as most beemen do every year or every other year if the queen is exceptionally good, the bees will supersede their old queen themselves, especially during the third year, when the old queen will likely begin to lay fewer eggs. Queens do not often survive more than two good years of egg laying.

Moreover, if a new queen happens to be deficient in egg-laying capacity, a colony may start building supersedure cells almost at once following her introduction. A queen deficient from the start may be the result of poor feeding during the larval stage or of overheating or chilling—or imperfect mating. Sometimes two queens, mother and daughter, will be found laying side by side on the same comb—to all appearances quite amicably; but this does not usually last long. The new queen will often sting the old one to death—the queen, by the way, being able to do this without losing her life.

All this is another reason for frequent inspection of your

colony. It may cast a swarm unless you take care of extra queen cells and add supers when needed. It is a good thing to be able to tell, also, when a young queen is ready to emerge. Here the other bees will help you; for, the day before she is to come out—or even thirty-six hours before—they begin to tear down the wax on the tip of the queen cell, as if to help her and give her less work to do in making her exit. Sometimes they do not do this, however; yet

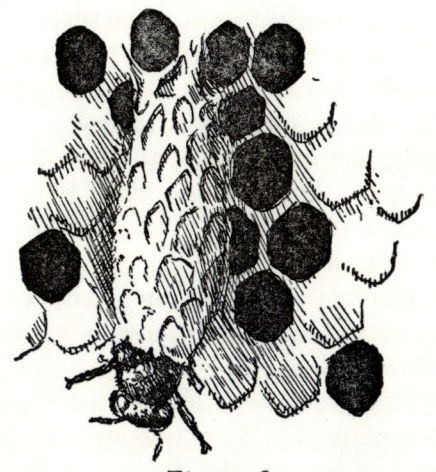

Figure 6
New Queen Emerging from Her Cell

the new queen has no difficulty in cutting her way out; so the reason for their action is not clear. If your queen cell is on new comb or comb foundation, perhaps you can see her moving about in her cell after the fourteenth day if you hold a strong light behind it. Also you may be able to hear her cutting her way clear, if you have good ears, and see her mandibles protruding from the cell as she cuts. All this is fun for the new beekeeper.

After the new queen pushes open the little door she has made, she looks for a cell of unsealed honey and takes a sip. Before mating, she usually finds her own food, but afterwards she turns to her workers for almost all her nourishment. She may dip her tongue into a cell of honey, but not often. The queen and the drones and occasionally another worker will dip their tongues into the mouth of a worker bee, who stands still, keeping her tongue folded behind her head.

After getting her first sip of honey, the queen crawls about and seeks the remaining queen cells, if any, to tear them down. If another queen has emerged before her, she discovers this, and a fight ensues. The first to sting the other survives this battle unhurt. It is usually the older who survives, as she is the stronger of the two.

In destroying the queen cells, the young queen merely cuts a hole in them, as a general thing, and leaves the rest of the destruction to the bees. If you want a queen for another colony, you may remove the extra cell or cells carefully before the one to be retained hatches, and attach them to a comb in the other colony, using a toothpick to fasten each one in place.

We cannot say that bees always act thus and so. It sometimes happens that two queens will live in quite friendly harmony for months, laying on the same comb even when they are sisters who have emerged at the same time. Mother and daughter may also work together for a long time, especially if the mother is old and not laying well. It even happens that a young queen will pay no attention to remaining queen cells but will wait until they hatch and then fight it out with the new queen, or else swarm with a lot of her bees. You can never be sure; so it is well to be on your guard and know what is going on in your colonies by frequent inspection.

The virgin queen, when she first leaves her cell, is a little smaller than the mated, fertile queen. Moreover, she decreases in size and by the fourth or fifth day can hardly be distinguished from the workers. It scarcely pays to hunt for her. The mating flight generally takes place from four to ten days after hatching, very commonly on the fifth day if the weather is good. As the young, newly hatched queen becomes stronger, she will begin to crawl about the entrance of the hive. When three or four days old, she will occasionally look out, as if inspecting the weather, and the following day, if the weather is good, will go out and try her wings. She is likely to do this during the warmest part of the day—a little after noon. She moves with the most scrupulous care, every motion graceful and cautious, and it is an enchanting sight to watch her. She seems to inspect the hive with the most careful scrutiny, taking wing for a few feet and then circling back as if to be sure that she knows her home. Sometimes she will not go out of sight at all for the first two or three flights, and then the next day, or later the same day, she will venture farther. Then comes the mating flight, when she may go out of sight, high up to the circle of waiting drones.

Occasionally a queen is lost on her mating flight. A bird may get her, or some other mishap befall. When she does return from a successful mating, she will still be small and insignificant in appearance, but a few hours before the first egg is laid—usually on the third or fourth day after mating—her body increases markedly in size and changes in color, becoming lighter. Moreover, she suddenly acquires a new dignity and ceases her nervous running about. She settles soberly down to her life work of laying eggs.

Queens commonly mate only once, although there are recorded instances of a second and even further mating.

If a queen does not begin to lay by the time she is twenty days old, she should be destroyed and superseded —unless it is out of season, when she may be held over to see whether she will lay when the right time comes.

You should know at once when a queen is lost, especially during the months when the honey flow is abundant. Even a single day lacking in egg production will mean the loss of 1500 bees, more or less, to bring in honey a month later. That means about one quart of bees! During this period, be extra careful in your inspection, lest you injure your queen, or do anything to hinder her in her work.

You will soon be able to tell a queenless colony by the way the bees behave. If they seem to stand aimlessly on the alighting board, it is a good idea to see what the matter is. If you find no newly laid eggs or worker brood, inspect to see whether you have a queen that is not laying, or no queen at all. If you do not find a laying queen, give the hive a frame of eggs from another colony, if you have one, so that they may build queen cells. If you have no other colony, try to get such a frame from a neighbor and order a laying queen at once.

After putting in a frame of eggs or brood, you should find the beginnings of queen cells twenty-four hours later, if the colony has been without a queen for a while. However, do not wait for these to be built and mature. Remove the frame and supply the new laying queen as soon as possible.

Another way of telling a queenless colony is by the buzzing of the bees. It is a distinctive kind of distress signal, begun as soon as a hive thus deficient is opened—very different from the ordinary buzzing. Then, if you find no young brood or eggs, you may be reasonably sure that the colony lacks its queen. Do not forget, however, that there may be queen cells even if your queen is actively laying; but the buzzing will be normal.

The same sort of abnormal buzzing may be set off by too much smoke—or, with some colonies, even a little. So, consider all the signs and what they point to, and act accordingly.

Usually you can depend upon the instinct for swarming or supersedure of the bees themselves to replace your old queen; but in the event of an accident or some unforeseen occurrence, or because you have a poor layer, you may wish to introduce a mature laying queen in the season of good honey flow. Again, you may want to requeen six weeks or more before the first heavy frost may be expected, in order to have a good supply of young bees before winter; this will help the colony to winter well and be ready for the spring build-up.

Sometimes the queen may get "cramps," that is, she will kink up into a crescent and drop down to the bottom of the hive, apparently lifeless. However, if she is let alone she will come out of the attack, most likely, and be as active as ever. Her retinue will be most gentle with her while she is in this condition, and wait around to clean her off and resume their care of her. We do not always know the cause of these cramps; they might result from handling or from fright.

4. *How to Requeen*

There are several ways of introducing a mature, mated queen. One is to place in the hive, in a special cage, a queen purchased from a good breeder. Ordinarily you cannot transfer a queen from one hive to another and give her freedom there immediately. She must be protected until the workers become accustomed to this "stepmother" and accept her. If you order a queen by mail, there will be accompanying directions for introducing her. First, these directions will tell you to be sure that your hive is queenless. If not, you must remove or destroy the old queen before putting the new one in. Introduce the cage with the wire front downward between two frames, over a cluster of brood, if there is any. Press the frames together to hold the cage in place. One compartment of the cage is filled with candy which is accessible to the bees through a hole in the end. They will gnaw away the small strip of blotting paper or cardboard which covers the candy, and you may daub a drop or two of honey on it to give them a start. Then close the hive and do not disturb the colony for about a week. The bees, gnawing away the paper, release the queen in about two days, more or less. If accepted, she should begin laying in a day or two, but she will not be wholly accepted until she has been laying for several days. So, if the colony is disturbed in the meantime, the workers may "ball" her and kill her. They do this by pulling at her and piling up on her in such numbers that they form a ball about her, and unless the operator comes to her rescue she will be suffocated or stung to death.

If the operator finds such a ball, he may lift it out of the hive and blow smoke on it lightly until the bees drop off one by one. He should always take care not to blow hot smoke on the queen. When she is almost released so that she can be picked up by the wings, the other bees may be pulled off her, and she may be caged again and given another try.

Another method of deballing which I have found very satisfactory is to fill a syringe with a thin syrup of honey and water, or sugar and water, and squirt this on the ball of bees until it is thoroughly penetrated. The ball may then be left in the hive. The bees will turn their attention to cleaning themselves and the queen, after which she will be accepted without question. This seems to give her the hive odor, and there is no further trouble.

After the two to two and a half days required by the bees to gnaw away the paper over the candy in the cage, another twenty-four hours is needed to eat out the candy—making from 72 to 144 hours in all. Then, if undisturbed, the queen will probably be accepted by the colony.

Another method of introducing is called the "Chantry principle." The cage here used is simple. It has a single opening of perforated zinc which gives the hive bees access to the queen and permits them to pass in and out. The bees begin to feed her and then mingle with the rest of the colony. This gives her the hive odor, actually accomplishing her introduction before she is released. There may be two cylinders of candy, a short one covered by the perforated zinc and a longer one which requires four or five days for the bees to consume. This cage, then, gives the bees admittance to the queen in one day and sets her free after about four days. There is no danger of the new queen being balled or killed while she is in the cage, and in the meantime the bees are being conditioned to her.

There are various modifications of these procedures, but for the beginner we have given the essential information.

5. *Requeening without Dequeening*

Sometimes you will want to requeen a colony without first getting rid of the old queen. This should be done only when the bees are actively bringing in nectar, but a month or so before the main honey flow begins. There are several ways of doing this, and here are two good ones:

1. At the peak of brood rearing, after putting the old queen below on the emptiest combs, put the brood combs in a second brood chamber above, on top of a combination bottom and cover. The entrance may be left at the front or put in the rear, as you think best. Give this upper brood nest a queen cell. When the queen hatches, allow her to mate. Then, when the heavy flow comes on, kill

the old queen and place the upper chamber with the new queen below on the bottom board. Put a newspaper on top of this chamber, with a few holes punched in it with a pencil; and place above this as many supers as desirable. Then set the old colony on top of all.

Inner covers, with three pieces of lath, will serve for the combination cover and bottom.

This method prevents swarming, and you have brood from two queens for the harvest. This is an especially good procedure when a late flow is expected.

2. You may permit the old queen to work in the brood chamber and in the food chamber, that is, in both lower and upper stories, during the months before the main honey flow begins; but when a better queen is desired because of a prospective heavy flow, you confine the old one with most of her brood to the lower story by using a queen excluder. Some well-filled combs of honey should be in the upper story, and if there is no brood there already, some should be transferred from the brood chamber: from three to five frames are enough. Then you should give a maturing queen cell to the super and set it off on a bottom board a few feet away from the parent hive and facing in the same direction. Put a hive cover on the brood chamber from which you have removed the super. Most of the old bees in the super will return to the main hive, but enough of the younger ones will remain to care for the brood and rear a new queen. A stock of queen cells may be provided in advance, or the queenless bees will rear a queen from the larvae in the cells you have given them.

When the honey flow is over and you have removed all that is not required by the bees for wintering, you may return the super with the young queen to the parent hive. If this takes place in the fall, you will not need a queen excluder (or the newspaper mentioned earlier) between the two stories. In most cases the young queen will be kept and the one in the brood chamber will be destroyed. Since the bees in the upper story must use the main hive entrance after being replaced, they will probably kill the old queen as they pass through her domain, whereas those already in the lower story are unlikely to enter the super; so the new queen is not disturbed. In one operation you have both requeened and provided a supply of winter stores for a strong colony in the spring.

6. *Swarming*

Swarming divides a colony and thus weakens it. It is the natural method of division, but the beekeeper always wishes to prevent it, lest his colonies lose strength for the honey flow. Swarm prevention should begin in the fall, before winter sets in. This is done by seeing that the hive has at least thirty pounds of honey for its use during the winter, and that no weak or diseased colony is allowed to continue. Every colony to be held over should have a good queen.

Then, in the spring, with your colony free from disease, with plenty of honey, and a good queen, you may leave it alone until the queen has just about filled up the brood chamber. By that time, the drones will be flying, and unless they are given more room the bees will swarm. So add a super of full depth, and later, if necessary, put on another, either shallow or full. Move some of the brood combs from the lower story into the upper, replacing them with empty combs supplied with full sheets of comb foundation. This should be done before the colony starts queen cells; but if the bees have got ahead of you and started this operation anyway, destroy the queen cells.

A colony is crowded when there is no more room for the queen to lay or for storing honey. As this period approaches, queen cells will be started so that the colony can swarm. Other factors that tend to induce swarming may be an old queen, or a light run of nectar just preceding the main flow; but crowded conditions are the most common cause.

When a swarm does get away, only about one third of the bees will have remained in the hive. The brood will be all sealed, and not require much attention. It will be hatching all the time, and within ten days or so will have gone a long way toward acquiring its original strength. However, if you have left more than one queen cell, the first virgin to emerge will lead off another swarm. Colonies have been known to cast swarm after swarm for several successive days, especially among Carniolans. So do not forget to destroy all the extra queen cells.

Another way of discouraging swarming is to have a double brood chamber. The bees will gradually eat out the honey and pollen in the upper food chamber, which will thus provide more room for the queen to operate upstairs when the lower story is filled with brood. The increasing room in the food chamber turns it into a brood chamber, and there may be as many as eighteen frames of brood at one time in such a hive. Additional supers will give room for honey storage. Later, when the peak of honey production has passed, the queen may be confined to the first story by a queen excluder.

How to hive a swarm will be told in the next chapter.

7. *A Few Pointers on Reducing Swarming*

Besides what you have already been told about swarm prevention, here are a few additional pointers:

1. Clipping the Queen. It is sometimes a good idea to clip a wing of your queen, because if the bees want to swarm this prevents their escape when you are away. Yet you may lose a queen if a swarm goes out during your absence: since she cannot fly, she may run out in some weeds and be unable to get back when the bees return—

A FEW POINTERS ON REDUCING SWARMING

and they will return if she is not with them. If they do return without her, you should destroy all but one of the queen cells which may be in the hive, and they will soon rear another. Or, if it is close to the honey flow, you should provide them with a new proved queen by the cage method.

If you fear injuring your queen by clipping a wing, practice first on a few drones. You may clasp the wings between the forefinger and thumb of your right hand (unless you are left-handed), and pick the bee up carefully. Then take gentle hold of head and thorax by the other hand, release the wings and, using a small pair of scissors, clip off the large top wing on one side. Some beekeepers clip off both wings, but this is unnecessary. Do not cut too close: a little more than halfway down the wing is sufficient. All the while be very careful not to pinch or restrict the abdomen in any way, as the queen might thus be irreparably damaged for egg laying.

2. Remove drone comb, as it tends to crowd the brood chamber and reduce profitable brood rearing. Remember that crowded conditions induce swarming!

3. Make sure that the hive is well ventilated. The entrance should be fully opened when the honey flow is on. If some bees want to cluster and remain inactive, even more than the normal ventilation should be provided if the weather is warm. Blocks may be inserted between the hive and the bottom board. With all this ventilation, the lazy bees will go back to work—if there is room.

4. Provide more super room at the least indication of crowded conditions.

5. In hot weather, give enough but not too much shade. White or aluminum hives are cooler than unpainted ones. If the hive is uncomfortably hot, some of the bees may swarm.

6. Demareeing. To demaree means to follow the De-

maree plan of swarm control. Mr. G. W. Demaree had several modifications of his plan, but this is the main outline: (a) Put the queen in new quarters, where she will have ample room to lay. (b) Place the sealed brood in an upper story away from the brood chamber below, and put a queen excluder between. (c) Give plenty of room for the field bees to store nectar.

When the emerging brood is all upstairs, it makes room, as it hatches, for storing honey. The brood below is gradually sealed and finally gives way to young bees. The upper story that had held the sealed and emerging brood will be filled with honey, and then more supers may be added. Mr. Demaree's plan thus produces in the hive the exact conditions that obtain when a colony swarms, while at the same time keeping the swarm and the parent colony together.

7. Have your colony or colonies strong. This is just good common sense in beekeeping, but it also reduces swarming, because a strong colony will readily expand into the supers. Additional supers should be added during the early part of the honey flow, when swarming is to be expected.

8. Remove one or two combs of brood from a lower brood chamber and place them in a super. If young bees emerge at the top of a hive, it gives better distribution throughout. Of course, you will not want to put brood in a super designed for comb honey.

These methods and precautions doubtless overlap, but the repetition emphasizes the points which are important.

8. *Laying Workers and Drone-laying Queens*

A situation which you are not likely to encounter with your one or two hives, if you inspect them often enough and care for them accordingly, is that of laying workers or of a queen which is a drone layer. If a colony has been neglected and is hopelessly queenless for several weeks, some of the workers will begin to lay. The eggs are abnormal in appearance, and there may be up to a dozen deposited in one cell. A few of the eggs may hatch, but the larvae usually die or produce only stunted drones. Queen cells may be started around some of them, but they do not develop a queen. Moreover, you cannot introduce a good laying queen into such a colony, nor can you induce them to accept a good queen cell. The best thing to do is to shake all the bees off the combs, give the combs to a normal colony, and then kill the bees with a charge of calcium cyanide after they have gone back into the frameless hive.

A drone-laying queen—one who lays only drones—is either one that has been imperfectly mated or an old queen who has used up her supply of fertilized worker eggs. It is generally thought that the drones from a young, infertile queen should be destroyed with their mother, even at a time when drones are necessary to the colony. The drones from an old queen who has been a good layer may be kept if they are needed. However, a good young queen should be provided as soon as possible.

9. *Bee Diseases and Their Treatment*

The two most common bee diseases are American foul brood and the European foul brood (AFB and EFB). The symptoms of AFB are these: the worker larvae die and decompose in the cell, both before and after being sealed. They give off a nasty glue-pot odor and settle to the bottom of the cell in a brownish, gluey mass that will string out from the end of a knife. Eventually they become almost black scales which stick to the lower side of the cells. The combs become mottled and greasy looking, showing sunken and perforated cappings. The bees do not make any progress in cleaning out the dead matter.

American foul brood is readily detected in its beginnings when the comb is held so that a light strikes the lower walls of the cells.

This disease forms spores, and these attach themselves in great numbers to honey, comb, and equipment that comes in contact with them. Thus it is spread by robber bees or by re-use of infected equipment. So, until recently, killing all the bees and burning the entire contents of the hive was considered the only method that would satisfactorily eradicate the disease. Calcium cyanide was usually employed in killing the bees, and the entire contents of the hive—frames, larvae, combs, and honey—were burned and buried. After this, the hive was thoroughly disinfected by scorching. If it was old, it was often burned, too.

If you get into trouble with this disease and wish to use the old method, the calcium cyanide may be placed on a card and inserted in the entrance of the hive. Sulphur fumes blown in the entrance by a smoker will also do the job. This operation should be performed after dark, to get all the field bees, and the burning should follow as soon as possible. If there is a lot of honey, additional wood may be required to burn it thoroughly. All should be burned to a cinder, and deeply buried. The hive, cover, and bottom board may be disinfected with a blowtorch. The wood should be charred. Gasoline may be applied to the inside of the hive and fired. Also disinfect all tools used with that hive; and, last of all, wash your hands with a strong disinfecting soap.

When American foul brood is found in many hives in a large apiary, the hives are usually disinfected by boiling in a deep tank for ten or fifteen minutes in a strong solution of one pound of lye to fifteen gallons of water.

In recent years many beekeepers have found sulfa drugs effective in curing bees with American or European foul brood; but unless it has been tried and found successful in your locality, it is best to stick to the disinfecting and burning method, especially if the disease has spread widely in your hive. While sulfathiazole may be fed to the bees in sugar syrup, it is probably better as a preventive of AFB than as a cure.

European foul brood is not so serious as the American in its effects, although it manifests some similar symptoms. EFB attacks both the worker and the drone larvae, especially during the unsealed state. Prevention is better than cure, and good beekeeping with strong colonies makes it difficult for EFB to get a good start. The affected larvae curl up in the cell and die in various positions which are retained until they dry up. A yellowish tint predominates in the dead larvae, although it deepens to a dark brown as the scaly stage is reached. The dead larvae, tested with a toothpick, lump rather than rope and are easily distinguished from the stringy larvae of AFB. They appear also

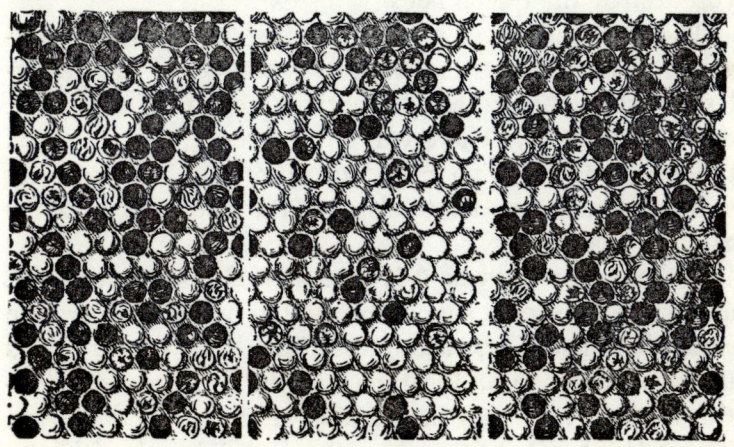

Figure 7

European Foul Brood *Normal Brood* *American Foul Brood*

to be more watery and may or may not be foul smelling. EFB spreads faster than AFB, both within the hive and from hive to hive. Apparently the bacillus of this disease is not a spore former and thus does not permanently infect the honey or the combs; nor does it exist long if deprived of its only food supply, young larvae. So the purpose of the treatment is to deprive it of its food. This is done by temporarily bringing brood rearing in the infected hive to a stop, killing the queen and requeening the colony from a good stock. If the attack is very mild, the queen may simply be caged for a week, instead of killed. If dequeened, the colony should be given a mature queen cell in about a week, as the bees may fail if the job of requeening is left to them. Queen cells in diseased colonies do not always hatch, and thus should be destroyed at the time the new queen cell is introduced. If you do not have access to mature cells from healthy stock, give the sick hive a brood comb containing eggs from a healthy colony when you destroy the queen cells.

The introduction of a young laying queen may be advisable, but in that event the colony should be left queenless for sixteen days, and all native queen cells destroyed on the eighth day. There will be an interval during which the bees will have no brood to nurse and consequently will turn their attention to cleaning the diseased combs for the new queen.

If the colony is weak, it is time wasted to treat for EFB, and such a colony should be united with another one similarly afflicted or, lacking that, should be destroyed, as in the case of AFB. Treatment in the fall practically always fails. The honey and combs need not be destroyed if care is taken to store all combs containing dead larvae until the latter are thoroughly dried. After this the frames may be used for super combs on disease-free colonies.

If no attempt is made to winter-over very weak colonies when EFB is prevalent; if the bees are fed in the spring to offset bad weather conditions, and if the hives are set at least six feet apart in the apiary, the disease is unlikely to spread. Never feed from honey syrup unless you are sure that the honey used is free from disease. Use equal parts of sugar and water, mixed when cold. Also, do not hesitate to call your bee inspector if something seems wrong which you do not understand.

Another bee disease, much less harmful, however, is nosema. This is a sort of dysentery, and while it may become very serious, it is usually checked by good beekeeping practices. Sometimes requeening with a healthy queen will clear up a colony, other conditions being good. As the spores of the disease are easily transmitted, the beekeeper should see that his colonies have a good supply of fresh, clean water and that stagnant pools nearby are eliminated. The antibiotic Fumagillin has been found effective in handling the disease, also.

These are the bee diseases most frequently found.

10. *Other Enemies*

In addition to disease, the bees have various enemies. Among them are birds, lice, skunks, mice, wax moths, ants, spiders, mosquito hawks, etc. Only queen breeders need be concerned about birds, and they should have their yards where there are no trees among which the birds may congregate. The depredations of lice may be serious, but fortunately they are found in only a few places and are easily killed by tobacco smoke, of which they can endure less than the bees can. Skunks eat many bees and greatly upset the equilibrium of the hive; they may cause the colony to dwindle when it should be building up. Skunks can be killed with rat poison.

To protect a hive from mice, a coarse wire mesh or hardware cloth that will let the bees pass through, but not mice, is sometimes fastened over the entrance.

Ants are not always troublesome, but there are some kinds that work havoc among bees. In the Southern states and occasionally in California, the Argentine ant attacks bees and brood and steals honey. Termites attack hive and bottom board, which they may riddle to a mere shell. Some ants will bite off wings and legs of bees, and a fight between ants and bees can go on until the bees are practically all dead. The best protection is vigilance. Watch the trail of ants going to your colony and find the nest and destroy it. Calcium cyanide may be used, or cyanogas. When I lived in Tahiti, I always kept my hives on legged stands, with the legs in cans of oil, so that ants could not climb into the hives.

Wax moths are sometimes a serious problem. There are several species whose larvae damage honey combs. Sometimes the colony has been previously weakened by AFB, and the dwindling of the colony as a result permits the wax moth larvae gradually to gain a foothold. A strong colony is usually able to prevent or forestall wax moth feeding, especially if it has proper bee spaces so that the bees can find and kill the larvae and carry them out. Stored combs not protected by bees should be protected by the beekeeper from wax moth invasion. Do not use DDT on them, as that would injure the bees when you put the frames back in a hive: just keep them in a place which the moths cannot enter. If they do get a foothold, the best thing for you to do, as a small beekeeper, is to burn the wax and scald the frames.

Spiders may catch bees in their webs; so just be sure to have no webs near your hives.

In some locations, mosquito hawks or dragon flies will

attack bees as well as mosquitoes. There is nothing you can do about it, and since you are not a commercial beekeeper it should not be a serious problem for you—unless you live in Florida!

CHAPTER FOUR

The Fundamental Principles and Practices of Beekeeping

---

1. The Three General Principles

RULES for good beekeeping are based on the usual behavior of bees, as outlined in the preceding chapter. However, each year will present its unusual problems. While bees of one race always react in the same way to the same conditions, it requires an expert student to recognize when and how these conditions vary and to fit his practice to any situations which may arise. One who may rightfully be called a bee master usually secures twice as much honey per colony as the average beekeeper in his vicinity; and, after all, honey getting is the reason for keeping bees—honey getting and pollination.

In my book *Honey Getting** I presented the three great commandments for success in producing honey:

1. Have each colony up to its maximum honey-storing strength at the beginning of the blooming period of the principal nectar-producing plants of the locality.

2. Keep each colony at this maximum point during the honey flow, not permitting the worker force to be divided by uncontrolled swarming.

3. Conserve the colony at all other seasons of the year, so that it may become strong at the right time.

These three principles are really one, as each emphasizes proper colony strength for the season. They deserve serious study. Rule-of-thumb methods, not based on principles, account for the frequent failure of beekeepers to secure a crop of honey when they move to a new location where conditions are different. Principles give a comprehensive view of beekeeping under all conditions.

2. The Eight Essentials of Practice

In carrying out these principles, eight essentials of good practice are necessary. I discussed these at length in *Honey Getting*. While that book was designed for the commercial beekeeper who wants to make money from his honey crop, the eight essentials may well be reviewed here, because the hobbyist and the amateur should know them too. Moreover, when that book was written, less was known than now about the help that supplemental heat can give to the beekeeper in some localities. We shall devote more thought to that subject later, but will touch upon it here in covering the eight essentials:

1. SECURE AND MAINTAIN A CLEAR BROOD NEST. This is the main thesis of all eight essentials of practice. By a clear brood nest is meant a brood chamber full of all-

---

* *Honey Getting*, 1947 (2nd Ed.). Earthmaster Publications, P.O. Box 488, Sun Valley, Calif.

worker combs and clear enough of honey and sealed brood to give the queen room to deposit as many eggs as she will. It will be necessary to use a queen excluder or some other effective method of controlling the location of the brood nest. Moreover, in cold spring weather, supplemental heat will enable the bees to maintain a temperature of 85° in the hive. This will encourage the rearing of vast numbers of young bees, so that enough of the right age may always be present to build good combs for replacing poor ones, and to bring in honey during the nectar flow.

2. KEEP EACH OF YOUR COLONIES APPROXIMATELY UNIFORM. While this is more important for the commercial beekeeper than for the amateur, it is well for you, too, to follow the practice if you have more than one colony. In the spring, your colonies should be in standard spring condition, and during the honey flow they should be of standard honey-storing strength. These terms are defined in *Honey Getting* as follows.

STANDARD SPRING CONDITION is defined as that combination of quality of queen, colony population, condition of combs, amount of stores, and all other natural and mechanical conditions which, with a standard method of management, will produce a colony of STANDARD HONEY-STORING STRENGTH at the beginning of the honey flow.

STANDARD HONEY-STORING STRENGTH means that a colony has such a population and is in such a condition that, under the system of management used by the operator, it will store an adequate amount of honey at low cost. Both STANDARD SPRING CONDITION and STANDARD HONEY-STORING STRENGTH will vary with local conditions and with the system of management used, and must be determined by each operator for himself.

In some cold localities, these conditions can be furnished more easily and to the increase of the beekeeper's profits by the use of additional heat; poor queens may then be weeded out more readily and optimum conditions secured for rearing new queens, for having good combs built, and for furnishing enough brood to share with weaker colonies—thus bringing all up to honey-storing strength at the right time.

3. KEEP ALL COLONIES QUEENRIGHT. A good queen is essential to a strong colony. Whenever the desired strength or uniformity is lacking because of a poorly laying queen, requeen as soon as possible. The use of heat in rearing good young queens early in a cold spring will be discussed in chapter 6. A young active queen helps to keep the brood nest clear of honey, since the bees readily remove honey to give her room to lay.

4. EMPLOY NUCLEI IN FULL-SIZED HIVES TO PROVIDE A STOCK OF QUEENS FOR REPLACEMENT AND INCREASE. Four-frame nucleus boxes may also be used. This is an essential practice for the commercial honey producer, but as an amateur you will not need to be very much concerned with it. Nevertheless, later in the chapter the meaning of nuclei and how to handle them will be explained.

5. KEEP RECORDS OF ALL LAYING QUEENS. This is easier to do if you clip their wings or practice some equally good method of controlling the queen and preventing swarming. This keeping of records is also for the commercial beekeeper.

6. VISIT AND WORK EVERY HIVE EVERY EIGHT OR TEN DAYS DURING THE SWARMING SEASON. By the use of pollen and supplementary heat, you may have assured a strong colony at the beginning of the honey flow, with the storing instinct dominant. Under such conditions, a colony will not swarm. Added heat in cold, wet weather will enable you to build up your colonies to strength and keep them that way. Correct hive temperature supplements good beekeeping methods. Then, to keep your colony strong,

you must know what is going on, and frequent inspection is necessary.

7. HANDLE YOUR COLONIES IN SUCH A WAY THAT THE BEES REMAIN GOOD-TEMPERED. How to do this has already been discussed. Sometimes supplemental heat helps here, too.

8. CONTROL AND ERADICATE ALL BEE DISEASES AND ENEMIES. All that you will require for this has already been told in chapter 3; but if you wish further information on this and other beekeeping essentials, read *Honey Getting* or some other good book for the commercial producer.

3. *The Clear Brood Nest*

The foregoing three principles and eight essentials of practice all center on having a clear brood nest for a good laying queen. A small colony, as the brood-rearing period comes on in a warm climate, may have a comb of sealed honey alongside its brood nest. If the weather is cold, the bees will remove this honey and keep the brood nest compact and protected, which is well. Otherwise, the queen will pass over the honey to an empty comb, laying patches here and there. This scatters brood throughout the hive, mixing it with sealed honey—evidence of poor control and very bad if you want to produce some nice comb honey. Even the outer surface of the outside comb may contain brood; yet such a colony may have only about six combs of brood all told, and thus fail to build up to honey-storing strength at the right time.

The best way to remedy the situation is to place one or two frames of comb foundation in the brood nest at one time, but no more than will be occupied by the queen as fast as the cells are drawn out by the young bees. Otherwise, there will be cells half drawn out and filled with honey and sealed. As is stated in *Honey Getting:* "Little can be accomplished by removing this sealed honey and replacing it by empty combs, because all cells except those immediately needed by the queen will be filled with incoming nectar."

In maintaining a clear brood nest the importance of good combs cannot be overemphasized. Have only good, all-worker combs in the brood chamber. With one hive body, eight combs, and with two stories, at least twelve combs should be kept in prime condition for the use of the queen. If more brood is produced than the colony needs, the commercial beekeeper may use the surplus to make his colonies uniform, or to build up nuclei.

If a comb has some pollen in it, the bees might divide the brood nest into two parts, one occupied by the queen and the other in exactly the right condition for building queen cells. Then you may expect swarming, and the situation is especially disastrous if you are working for comb honey.

When preparations for swarming are discovered, one or two combs of mostly sealed brood should be removed at a time and replaced by comb foundation, but not by empty combs, as these would probably be filled with honey without an egg being laid. As I have said in a previous chapter, the sealed brood may be put in a super, with a queen excluder between.

If you cannot inspect your hives every eight or ten days, almost all the brood may be separated from the queen in one operation, as in demareeing (page 59), and then the queen will establish a new brood nest. However, it is usually preferable to permit the unsealed brood to remain in the chamber where there are plenty of young nurse bees to care for it properly, or some of it may die. Moreover, if unsealed brood is taken away and only sealed brood is left, swarming is invited, because the colony is thrown out of balance; whereas, if sealed brood only is removed, the emerging bees are distributed and swarming is checked.

Some strains of bees are more likely to allow the brood nest to become honey bound than others. When necessary to keep the brood nest clear, combs of honey should be removed and replaced by foundation.

4. Maintaining the Colony in Strength

When the nectar-producing season is of long duration, even young and active queens seem to become less active and breeding slows down. Sometimes a Southern beekeeper, in an area where one nectar-bearing crop rapidly succeeds another, finds that his bees require two or three months to build up to a population which is the normal increase in the North during the four to six weeks preceding the clover harvest. This slow build-up makes it difficult to maintain a clear brood nest, as the combs become full of sealed honey and brood in scattered patches. Some honey may be stored in the supers, but the maximum amount will not be stored even from rich sources of early nectar because the bees will be breeding up on it.

The problem then will be to get the sealed honey and brood out of the brood chamber and give the queen ample room for laying. This honey may be extracted at once or, with the commercial beekeeper, put up in a super for later extraction after the brood has emerged. In warm climates, it is exceptional to have a queen quickly fill empty combs with eggs: too often brood chamber combs are filled again with honey even when there is plenty of room in the supers. The way to avoid this is to have the colonies strong in the early spring, before the long honey flow starts. Supplemental heat—bee weather supplied earlier than even the Southern sun can give it—is of great help in such an early build-up. Weak colonies are too common in the Southern sections of the United States because beekeepers give them less careful attention, assuming they can take better care of themselves. Unfortunately the assumption is not correct; and wintering in the South may be more difficult than in the North, where more care is taken to have strong colonies in the fall which, properly protected, will be strong colonies without special handling when spring comes. There should be a large population of young bees before brood rearing is discontinued in the fall. So—*keep your brood nest clear.*

5. Making Nuclei and Dividing

A nucleus is a very small colony with a queen. It may include only about a hundred bees, but such a small number cannot survive without assistance from the operator. Usually there should be 500 or 1000 bees, with one or two full-sized frames. When there are as many as five or six frames of brood and bees, it is thought of as a weak colony rather than a nucleus.

Making two- or three-frame nuclei for the purpose of increase is simple. First the queen of the colony from which the nucleus is to be taken should be located and the comb she is on should be set aside. Then put two combs of sealed brood and one comb of honey with the adhering bees into an empty hive: either a full-sized one or one just large enough to hold three frames—perhaps a full-sized one made smaller by a dividing-board. The entrance to the new hive should be sealed until nightfall. Then shake in the bees from two more combs. Introduce a new young queen at once by the cage method, and carry the nucleus to a new place. The next day some of the old bees will return to the parent stand, to which the old queen has been returned, but most of the young ones will remain in the new hive with the young queen.

Such nuclei should be made only in the spring so that they will have time to build combs and brood and to gain strength for the honey flow when it comes. Two nuclei, or even three, may be taken from a large, strong colony without seriously affecting its strength. Indeed, this is one method of preventing swarming. If there is not enough honey coming in, the nuclei should be fed until the flow commences: this is the reason for including a frame of honey.

The method of dividing a strong colony for the purpose of increase is very similar to that of forming nuclei. However, this is a wasteful practice, especially for the beginner, unless he does it at a time which will allow both colonies to build up to full strength for the honey flow, namely, six or eight weeks prior thereto. Then two or three young queens working busily may produce enough bees to bring in much more honey than the parent colony, no matter how strong, could have produced with only one queen.

6. Uniting Nuclei and Weak Colonies

Uniting colonies is just the opposite of dividing, but it is sometimes a useful practice, especially in the fall. It may be essentially the process of requeening without dequeening (page 55). It makes a strong colony for wintering, as a large number of bees can more easily keep themselves warm than a small number. Moreover, there is less likely to be "spring dwindling."

Yet, when two or more families of bees are put together that have had no previous association (as the two families in the requeening process have had), they may or may not accept each other. If they start a battle, the results may be disastrous to one or both families. Smoke is required to stop them in this eventuality. However, if the bees are from the gentler races there may be no quarreling, even if no smoke is used.

The uniting should be done in cooler weather when brood rearing has ceased and they begin to form clusters to keep themselves warm. After putting the colonies together, you should return in a few minutes to see whether all has gone well. Of course, the entrance is blocked, and if the bees are reasonably quiet, no smoke—or no additional smoke—is required. That night the entrance may be opened. Some of the old bees may return to the old locations, but if there is nothing there they will soon die or orient themselves to the new place. Of course, colonies so united should be free from disease.

Another way of uniting two colonies, the one which may be employed at any time, is to place a newspaper between the two hives. In hot weather, a hole or several holes should be punched in the paper. A single thickness is sufficient. The front entrance is blocked for 36 hours, and by the time the bees in the upper hive gnaw through the paper, they are indisposed to fight. If each of the united colonies has its queen and there is no choice between them, let the bees take care of the situation. If you know one is better than the other, kill the poorer queen and cage the better, and place her among the bees that are not her own, preferably the upper colony. The reason for this is that, when released, she will make her home there—at first, anyway—and will perhaps attract some of her own bees to move back and forth between lower and upper story, thus more quickly integrating the colony odors.

Colonies in pairs may sometimes be united easily. If two weak colonies are close together, one hive may be removed and the other placed in a spot halfway between the places where the two were originally located. Returning field bees from both colonies will enter this hive because it is so near the accustomed spot. At night the other hive may be placed on top of this one. If a newspaper is used between the two, the peaceful uniting of the colonies is assured.

These are practices which you will be unlikely to need during your first year as a beekeeper; but you should know about them, particularly in an emergency. Remember especially that uniting two weak colonies in the spring is wasteful and unprofitable but is a good practice for the fall, to insure wintering safely. Uniting a weak to a strong colony in the spring may be quite worth while, as it will have a strong queen.

7. Hiving a Swarm

Two swarms of bees may be put at once into a new hive together, and they will accept each other without question, although one of the queens will soon be killed. You may take care of the extra queen yourself, as she will easily be found not long after the colonies are united, because the bees will have begun to ball her.

Swarming divides a colony and thus weakens it, as we have previously emphasized. It is the natural method of division, but the beekeeper always wishes to prevent it, lest his colonies lose strength for the honey flow. Nevertheless, in spite of your precautions, a swarm might get away from you; or you might find a swarm in the neighborhood which you would like to have. Here is the way to go about hiving it:

Approximately two thirds of the bees in a hive join the swarm, and the queen is usually the last to leave. They do not go far away, as a general thing, but will soon settle on a tree or shrub in the vicinity. They should be nicely settled before you attempt to hive them. Prepare a hive with combs and foundation—it is even a good idea to put a frame of young brood in the hive, if you can spare one; and place this close to the swarm. If they are in or near the top of a tall tree you may find them difficult to capture. In that event you will need a stepladder and a bushel basket or a large dishpan. You may either saw off a limb carefully, in such a way that you can get hold of it and shake the bees into the container; or, if possible, you may jar the limb so that the bees will drop into it. If you can drop them directly into the hive, so much the better, but if they are up high, a basket or pan or, better yet, a cheesecloth bag with an iron or wire hoop about twenty inches in diameter at the open end may be used to catch and transport them to the hive. If such a bag is attached to a pole, it can sometimes be drawn up around a swarm, which is then shaken into it. Your own ingenuity must often be called upon in hiving a swarm, as it would be impossible to give directions to cover every contingency.

I knew a boy once who found a swarm in the woods over a mile from his home. He was an enthusiastic young beekeeper and he wanted those bees, although he had no container with him. He shook them down all over his body, where they settled as happily as on the limb where he found them, and he walked home with bees clinging to hat, face, and body. He prepared a hive, brushed the bees off himself with a turkey wing—and soon had a thriving new colony! He received only two stings! The bees felt no odor of fear about him and did not become disturbed.*

Once the swarm is captured—in whatever container you use—the bees may be dumped in front of the hive and they will soon enter and set happily to work. If some more bees cluster back on the spot where you discovered them, the operation may be repeated if it is worth while.

If your swarm is a long way from home and you must transport the bees from some distance, a bag is the best

---

* This is evidently not a completely unique experience. My friend Allen Latham, in his *Bee Book,* tells of doing the same thing when he was a boy!

device. It can be tied around the swarm and left a while until any bees that might be flying have settled on the outside. The limb can be severed with pruning shears or a

Figure 8

Catching a Swarm in a Looped Bag

saw and carefully transported by wheelbarrow or truck.

After your swarm has entered its new home, place the cover on loosely, to allow a current of fresh air to flow. When a full day has passed, the cover may be settled tightly in place.

8. Preparing Hives for the Honey Flow

Before the honey flow begins, you will want to have your supers and frames ready to add to your hive bodies. For straining or extracting your honey, it is possible to use old frames, if they are free from disease and you have scraped them as clean as possible with your hive tool and have added new comb foundation. If you want comb honey, the frames would better be new; this is mandatory if you want to produce fine section comb honey.

In any case, you will secure your supers and new frames (both unassembled) and comb foundation from some good bee supply house. You will also want queen excluders, one for each colony. For best results, and for rapid and easy handling, your frames should be of the Hoffman, or self-spacing, type (page 27). I prefer those which are 1⅜ inch from center to center; evidence points to this as being the natural bee space; but some beekeepers like a space of 1½ inch: it does permit a little less careful handling!

With your new supers and frames will come complete directions for assembling. There will also be instructions for wiring the frames and for putting in the comb foundation. Sometimes the comb foundation comes already wired; this is for strength and is especially useful in hot weather, as the honeycomb is less likely to sag from the heat and the frames may be handled more rapidly and less carefully without damage. The small section boxes for sectional comb honey are not wired, but for bulk comb honey they may be or not, according to your preference. If specially spaced, the wires may be avoided when you are cutting out the honey; and if you expect to take off honey on a very hot day, they do give additional strength.

If your hives are prepared, if you have observed precautions for preventing swarming and have made frequent inspections looking toward the maintenance of a *clear brood nest,* and finally if you have a good queen, you will have a strong, honey-producing colony in any area which furnishes a good nectar flow.

CHAPTER FIVE

Harvesting a Good Crop of Honey

THE three fundamental principles and the eight essentials of management outlined in chapter 4 all have in view a *clear brood nest;* a clear brood nest with a good queen means a strong, honey-producing colony. With good combs, work in the supers will begin promptly with the honey flow and will be continued without interruption as long as the bees can bring in nectar and pollen. Bees that die while the honey flow is on will be continually replaced; and at the end of the season the colonies will be so strong that they will winter well, with the proper precautions, and be ready for the next season in the desired condition.

1. Getting a Good Crop

Remember: nothing is more important than to have brood combs that are as perfect as possible. The brood nest must be as large as is required for the queen to lay all the eggs needed to build up the colony to standard honey-storing strength at the right time, neither too early nor too late; and it must be directly under the supers in which honey is to be stored.

If you want especially to produce comb honey instead of the extracted product, the use of a double or two-story brood chamber, with clear brood nests, is excellent. Until the time comes to add on a super, put no restriction whatever on the room for brood rearing. The double brood chamber furnishes more space for the combs and so discourages swarming, and it gives more room for storing pollen, thus keeping it away from the comb-honey sections. It also avoids extra handling.

The brood nest should extend from side to side of the brood chamber, directly beneath the super, and the brood should reach to the top and bottom bars of the frames; there should be no space in which honey can be stored without being easily removable. The brood should also extend all the way across the brood chamber, with no strip of sealed honey above the brood to keep the bees from starting work in the super as soon as it is added, and continuing to store honey as more supers are supplied.

In parts of the country where the main honey crop is from clover, the honey flow begins about ten days after the first blossoms are visible. Then is the time to put on a comb-honey super; another super may be added when the last one is about half full.

Even if you are using the deep Dadant hive and frames, it is important to keep the brood chamber so full of brood that there is no room for storing honey there. And for comb honey it is especially important. It requires particular care and attention to details.

You may have either "bulk comb honey" in shallow frames or "section comb honey" in the familiar squares. Nowadays, cellophane for wrapping the honey when cut out from the shallow frames, and cartons of the right size for holding the wrapped cuts render it less important to go to the extra trouble of producing it in the prepared sections. Squares of the same size may be cut from the frames, and when these are drained and wrapped, they are very similar in appearance.

For the bulk comb honey you will want shallow supers with special frames. Fasten the foundation in with melted wax. This is sufficient if you want the product for your own family or immediate neighbors. If you expect to transport it for some distance it is safer to cut out the corners of the top bars in the usual way and nail in the foundation to be waxed afterwards. You will not want sagged or warped combs if you desire fine bulk comb honey, and you may even prefer to wire your frames to measurement if the weather is going to be very hot when you harvest the crop.

The foundation should be put into the frames not long before it is needed, as the bees will start work more promptly and finish the combs more smoothly the fresher it is. It should extend only to within half an inch of the bottom bar, and to a quarter or an eighth of an inch from the end bars. Then it will hang perfectly straight (if your hives are level, as they should be) and will be fastened only lightly to the end and bottom bars—which makes cutting out easier.

If you can, plan your comb-honey production after the full flow of white honey has started. Place a queen excluder on the colony (unless you know that you can get satisfactory results without doing so), and on it place two to four of your special shallow supers, as they are needed, supplied with frames and comb foundation. If you set off a full-depth super pretty well filled with honey, you may put it back above the comb-honey supers, with a bee escape between, to ensure that any leakage is cleaned up before the honey is taken from the hives. When the bees have left the super or supers above the escape, they may be removed. All the comb-honey supers should be off the hive too before the honey flow ends, and may be removed as soon as they are filled. Then there will be no unfinished combs.

Be careful not to crowd the bees too much—just enough to start them on the foundation. Then remove the food chamber or extracting super, as I have outlined, and give the colony as many more supers as it requires, so that the combs will not be built too thick or fastened with burr combs to the frames below.

As soon as your colonies have finished their sets of comb-honey supers, they may be returned to extracted honey.

Of course, when the comb-honey supers are put on, the brood nest should be clear—in the best possible state for the prevention of swarming and for storing honey in the supers. The chamber should be well filled with brood, not unsealed larvae. If there are any full combs of honey in the brood chamber, they should be taken out and replaced with frames of brood from another hive, if possible, or with foundation, but not with empty combs. A single comb in the wrong place may induce swarming and cost you your crop.

Examine your hive about ten days after putting on the comb-honey supers; look for queen cells and see that the work is going well. If any colony is not doing well with comb honey, the commercial beekeeper will return it to working for extracted honey immediately, and will transfer the comb supers to colonies that are doing good work on comb honey and need room. He should not be tempted to put them on new colonies and so confuse the working out of his plan. He would not want to increase his cost of production by doing special work on a few colonies.

With you, however, the special work is justified, and you might keep your colony working on comb honey even if it should swarm.

The principles of production are exactly the same if you are producing section honey. The advantage of this is that the honey does not require wrapping. During the winter prepare as many section supers as you expect to need. Use only clean, straight, new separators, or you may find yourself hampered by brace combs. Separators are little fences consisting of a strip of wood or metal a little smaller in width than the height of the sections and equal in length to four sections standing side by side. They are alternated between the several rows of sections and prevent the bees from "bulging" the comb from one section to another. Without them, the sections would be uneven.

Again put in the sections and foundation shortly before the time to use them. Fill the sections with thin surplus foundation. Full sheets must be used to get sections of the finest appearance, but you may use only bottom starters if you do not care greatly about the appearance of your sections. Also, place the foundation in the sections with the cut edge attached to the wood. This puts the rows of cells crosswise, and gives the comb a much better appearance than if they run vertically.

A commercial comb-honey producer will take care to wedge or clamp the sections, filled with foundation, perfectly square, so that they will ship well, and you will do the same if you want really good-looking honey.

When you use smoke to drive the bees out of supers, you must be sure to have clean fuel in your smoker, so that ashes or soot will not be blown over the combs and into cells partly filled with honey. Do nothing to mar the appearance of your product, if you have pride in your work.

2. Taking Off Honey

Occasionally an amateur beekeeper is so eager to have some honey of his own production that he is tempted to take it off the hives before it is fully ripened. *Do not do it!* When it is first stored by the bees, honey is thin and watery and has not the fine flavors which are developed as it evaporates and is chemically changed and sealed over by the bees. Most nectar is largely sucrose, but it changes to levulose and dextrose as it becomes honey. When allowed to stay in the hive or in the comb for some time after it is sealed, it acquires a richness of quality which unripe honey does not possess. Moreover, it does not deteriorate and ferment as unripe honey does if removed from the hive.

Then too, some unripe honeys have disagreeable odors and flavors. Basswood, onion, and goldenrod honeys are all disagreeable when unripe, but this feature almost or entirely disappears if they are allowed to mature. It is like the flavor of wild onions in milk which persists if cows are milked too soon after coming from pasture but which disappears if milking is delayed.

Good ripe honey should run to not less than 11¾ pounds per gallon. As an amateur, until you have read more than this book tells you about granulation, fermentation, and the heating of honey, you should not attempt to market any of your product except among family or neighbors who will use it up quickly.

To remove your honey from filled supers, it would be best for you to use a bee escape, although there is another method employed by commercial producers for comb honey—the carbolic acid plan—which you may read about in any standard reference work on bees. The bee escape (see also page 85) is a device that permits the bees to go through a one-way passage, so that they cannot return into the super by the way they went out. There are various types on the market, and you may select the kind that most appeals to you as described in your catalogues. To insert the escape board, first loosen the super with your hive tool so that the propolis which glues it to the brood chamber is broken. Tilt the super up at one end with one hand while with the other you blow a few puffs of smoke in to make the bees retire. Then lift the end a little more until it is at an angle of about 45° with the hive. Set your smoker down and lift up the escape board, being careful to have it right side up. Slide it gently along the hive as far as it will go, with the bee-space side on top, and let the super down slowly, lastly bringing hive, escape board, and super into alignment.

If this is done before the honey flow has begun seriously to diminish, the bees above the escape will not be likely to cut open any of the cappings and fill themselves with honey, as they may do if escapes are put on when the honey flow is poor. The use of a minimum amount of smoke also helps to avoid this cutting.

Put your escapes on in the morning. During the day the field bees will leave, and they cannot get back. The next morning the supers will be ready to come off with but few bees in them.

Of course, you may instead simply smoke your bees off the frames and shake them—if you want to. However, you may bring too many bees into the house or extracting shed with the frames, and they are a nuisance, especially if you have several hives. Also, you will not want to do much shaking of comb-honey frames, lest you injure the appearance of the honey comb, a thing very easy to do on a hot day when the wax is soft.

Remember to have your smoker full of clean fuel.

Sometimes the bees are slow about sealing the honey, and you may have to tier up your supers to as many as six before any are so finished that they may be taken off.

However, toward the end of the honey flow, they will be sealing the cells more promptly. If you want nice-looking comb honey, it is well not to leave the supers on until all cells are sealed, especially if you expect to market some of your product. The wood will become more soiled with propolis, and the honey will assume a worn appearance if left too long. Experience will very soon teach you when it is the best time to take off your comb honey. Unfinished sections may be added to your honey for straining or extracting, or you may consume them promptly at home.

Since you will have your colony crowded a little for super room, most of the combs will be finished even in the outside sections. The number of supers should be reduced to one or two as soon as possible, and any unfinished frames or sections may be placed in these. With bulk comb honey it is easy to transfer the frames, putting the unfinished ones toward the center to encourage sealing. This practice at this season, with the honey flow almost over, is not likely to encourage swarming, especially with a lot of old bees in the hive.

3. *Caring for Your Comb Honey*

In removing the filled sections from supers, you must be careful to loosen them without twisting the frames; this will damage the combs by slightly cracking them and will start leakage. In producing bulk comb honey, you will carefully cut the combs out of the shallow frames into sections of equal size and set them on a suitable rack for draining out the cut cells. Three or four sections may be cut out of one shallow frame, depending on the condition of the comb around the edges. After being thoroughly drained, they may be neatly wrapped in cellophane.

Comb honey should not be kept too long, as it may

start to granulate very soon. To postpone granulation, it should be kept in a warm room, 80° to 85°.

If you are producing finished sections, you should scrape off all the propolis with a case knife or a jackknife to give them a neat and pleasing appearance. Or you may paste a strip of No. 2 sandpaper down on a table and rub the edges back and forth on the surface of this. This is not practicable on a warm day, or if the propolis is soft. Moreover, it is possible to damage the surface of the comb with the dust of the scraping.

You may use rubbing alcohol to remove the propolis from your hands. Gasoline or turpentine will do, too, but be careful of the odor around your nice honey.

4. Strained Honey

Except for your comb honey, which requires special care in handling, the manner of harvesting your crop need give you, as a small beekeeper, little concern. If you have one or two hives only, it will be a very simple matter to follow the old-fashioned method of processing your liquid honey. When it is characterized as "strained," nowadays, we usually refer to honey pressed and strained from the combs. Until about 1870, all liquid honey was so produced, and it was frequently contaminated by sediment consisting of wax, pollen, propolis, bees' remains, etc. There is still some ambiguity as to the use of the word "strained," because extracted honey is filtered through a strainer, too, and often the word describes this process. Strictly speaking, however, strained honey is that produced by the old method; it is now universally discarded by commercial beekeepers but still used by those who have only a few colonies of bees.

To produce liquid honey by this process, you need simply cut it out of the frames in small chunks and allow it to drain in a large pan—an enameled dishpan will do very well. You may crush the chunks with anything suitable—my wife has used a potato masher—and let the honey warm in the sun (if protected from bees!) or in any very warm place. Then it may be poured through two thicknesses of cheesecloth to remove the wax and other extraneous matter. Place the cheesecloth, clean and wrung out of clean warm water, down near the bottom of a suitable container—I like an enameled pail for this operation—and have it large enough to hang over the sides. Pour the

Figure 9
Uncapping for the Extractor

mashed comb and honey into this and gradually lift up the cloth, tie it, and hang it above the receptacle to drain, like jelly. Warm honey will flow much more readily than cold honey. So keep the room warm; but bear in mind also that honey which is heated too much darkens and loses its delicate flavor. It should never be heated to more than 160° F. and not to more than 140° when there is wax in it which shouldn't be melted, as in comb honey.

More honey may be coaxed out of the comb by putting it in a porous bag stronger than cheesecloth, and placing it under pressure.

5. *Extracted Honey*

If you eventually have as many as four or five hives, you may want to have an extractor, which removes honey from the combs by centrifugal force. The combs are first uncapped by a steam or electric uncapping knife, or by one which is dipped into hot water to allow it to slip easily through the cappings. This last will be quite adequate for you. The knife should never be hotter than 145°, the approximate melting point of wax, lest a burnt flavor be added to the honey.

A homemade uncapping outfit may consist of a wash boiler, supplied with a wooden support across the top on which the frames may rest as you uncap them; a primus, or small lamp stove on which to keep the hot water for the knife; and your uncapping knife—or, if you have many frames to work at once, two such knives, so that one can be heating while the other is in use.

Notch the stick across the boiler so that it will not slip when you rest a frame on it for uncapping. Drive a nail in the center, and balance the combs against this as you uncap them. The frame should be tipped forward, and most operators find it better to cut upward rather than downward.

When your knife needs sharpening, do not use a whetstone: it makes the edge too fine. Use a scythe stone, not too coarse, to keep the saw edge sharp. There is a stainless steel knife on the market with fine teeth which rarely, if ever, requires sharpening.

A small extractor usually has to be anchored to the floor, as it tends to slip during the extracting, especially when thrown out of balance by combs of unequal weight. It will have to be high enough for the honey to flow into the pail below. Directions for operating will come with your machine, if you buy a new one, but we review them here for your instruction if you purchase one second-hand.

The combs put into the extractor should be as uniform in weight as possible. It is well to extract at the same time old combs that are full of honey and not mix them with new combs only partially filled. Also, do not put a comb containing a lot of pollen in with others, as the pollen remains in the comb and throws the reel out of balance when the honey is removed from the other frames.

If you are operating your extractor by hand, start slowly and turn at a slow speed until about half the honey is thrown out of one side of the combs. Then stop the extractor and reverse the frames. Make another slow start, but this time gradually accelerate until you have reached your maximum speed. Get the second side of the combs as clean as possible and reverse them again, throwing out the rest of the honey on the first side.

Have a tank or large container adequate for the amount of honey you are handling, and put your two thicknesses of wet cheesecloth into this. Again the honey may be strained more easily and at a little cooler temperature if the cheesecloth is large enough to form a bag that goes almost to the bottom of the container. Then, as in the case of your strained honey, the bag may be lifted up and hung from a hook over the tank and allowed to drip down into it.

The advantage of extracting rather than simply straining is that the honey is comparatively free from impurities and passes readily through the cheesecloth filter.

Strain the honey from the extractor first, and then, if you have a two-frame extractor, from the cappings. If you have the usual three-frame extractor, you may lift out the three supports for holding the combs and dump in the cappings and the adhering honey. Capping wax is the best wax, and you should save it and melt it.

If you have a two-frame extractor, procure instead of the washboiler arrangement an uncapping can of standard design, consisting of a tub with an inner, smaller one of perforated aluminum or a basket of tinned wire, through the sides and bottom of which the honey can drain from the cappings. This has a wooden crossarm and a one-inch square bar long enough to extend across the outer tub, and a nail point extending up about an inch through the crossarm.

A four-frame hand-operated extractor leaves a greater residue of honey in the cells than does a two- or three-frame machine, for the reason that the hand and arm grow tired sooner in operating it; and all these leave much more than does a power-operated machine. Of course, a three- or four-frame electrically driven extractor is preferable. You will have better results more easily gained.

But even a small three-frame extractor, hand-operated (which is the one I recommend), with the rest of the outfit, would take care of more honey than that produced by your colonies. In fact, it would take care of an apiary of from 25 to 50 hives.

6. Care of the Extracting Room

After extracting, put the combs back in the supers, and to prevent dripping on the floor place these over something to catch the drip: if you have nothing better, several layers of old newspapers will do. If you are extracting in Mother's kitchen, you had better be clean, or else! But it is well to be clean wherever you are. A smeared floor in the extracting room or shed will attract bees, and if the operator carries smears on his shoes as he goes out, he may induce robbing.

If you are extracting at night in order to avoid robbing, return the empty combs to the hives the next day—in any case, as soon as possible. Even if the season is over, or nearly over, the bees will clean up the wet combs if you put them back in the hives; and when the combs are clean, they may be removed again and stored for the winter. If you have only a few hives, and there are no others near to do robbing, the wet frames may be put out in the open air after extracting and exposed to robbing by your own bees, who will soon clean them up; but this should not be done under any other circumstances.

If you have only one set of extracting combs to the hive, you will *have* to replace them as soon as they are empty. In this event, you should do your extracting on a day when the bees are flying and the weather is warm and pleasant.

7. Varying Characteristics of Honey

Some honey is so thick that it cannot be thrown out of the combs by the extractor unless first heated to about 100°. Your extracting room will probably not be as warm as this, and if you have any honey of this character, cut up the best of it, if possible, as comb-honey chunks, and mash the worst to drain in a very warm place, perhaps in the hot sun, covered well, or on the back of an old-fashioned wood or coal kitchen range. It may be necessary to apply pressure to get it all out of the combs. Heather honey is of this type.

After straining or extracting, your honey should be sealed in bottles or well-tinned cans and kept until required for table or kitchen. If it granulates, as most honey will, and some very quickly, it may be melted by setting bottle or tin in water, heated to not more than 160°. To avoid having it hot for too long, it may be poured off at intervals into another container. Honey should be quickly heated and quickly cooled to prevent discoloring or impairing the flavor.

Some honey granulates with a very fine grain and may be sold as "creamed" honey. This is excellent for the table, as it spreads nicely without dripping or "running" and is easier to handle than liquid honey on that account. If you have a honey which granulates with a coarse grain, you may make it fine by running it through a small hand grinder once or twice, sometimes adding a little liquid honey, if it is hard as well as coarse. Or you may induce it to granulate with a fine grain by first heating to 160°—to destroy any coarse granules—and then "seeding" or "priming" it with some dextrose purchased at the drugstore, or with a naturally finely grained honey. The amateur should not try to granulate more than about a pint in one container. After seeding, it should be kept at a temperature of 56° for about a week, more or less, as indicated by the results.

After bottling your honey and sealing it well, or storing the combs in a warm, dry cupboard, be sure that your extracting room is well cleaned of all scraps of honey and wax. It is a good idea to have a bucket or two of water and some suitable scrubbing cloths handy for this operation.

8. Rendering Wax

I have said that cappings are the best wax of all, and I have suggested that you might like to melt it down and make use of it. Here is the way to do it:

The cappings have been strained or whirled as dry as possible. If you have extracted fifty pounds of honey, you may have over a pound of cappings, although you will get less than a pound of wax. First wash the cappings well in warm water, or soak them for a day or so in rain or distilled water, and then rinse off. After the remains of honey seem to be thoroughly removed, put the cappings in a pan of hot water over a low flame. If you must use hard water, add about two tablespoonfuls of vinegar to a gallon of water, to prevent the wax from partially saponifying. Simmer at very low heat until the wax is thoroughly melted and risen to the top. The melting point of beeswax is between 143° and 147° F. (roughly 62° to 64° C.)—the highest of any wax known.

BE VERY CAREFUL IN HANDLING WAX AROUND A FIRE. Do not allow it to boil over on an open flame. Wax is highly inflammable and some serious fires have resulted from carelessness in handling it.

The pan in which you boil your cappings should be of enamel (agate) ware, or well-tinned, or of stainless steel or aluminum; iron or copper will discolor the wax.

As it simmers and melts, the finer impurities will rise to the top of the wax, and the coarser will sink to the bottom of the water. The top scum may easily be skimmed off by coaxing it to the center of the pan when you have turned off the heat. Some beemen, when they are rendering wax in large quantities, do this by dunking into the melted wax a cloth bag which just fits around the inside of the boiler, drawing it up all around the sides and gradually working the scum to the center as they lift up the bag. It takes a bit of practice to do this gracefully, but the scum can be drawn off in its entirety in this way if it is done at just the right time—a matter of experience.

Allow the wax and water to cool, then pour off the water, leaving your crude cake of wax. Clean this by scraping the bottom. If you wish to mold it into small cakes, melt it again, this time in a small pan set in a larger one of boiling water—or, perhaps, a double boiler. When soft enough, pour into small jelly molds or other suitable forms, which should be flared for easy removal of the wax. These little cakes make nice presents for mother or girl friend, to be used in ironing, waxing thread, etc.

Also, you may make a fine floor polish by stirring in some turpentine while the wax is still warm and soft, the amount depending upon the consistency you desire.

If your wax is from old combs, contains burr comb and propolis, and is discolored by age, don't simmer it loose but rather put it in a porous bag (bleached gunny sacking is good) and have some device ready for holding it down under the water. Hold it under hot, almost boiling water for some time and keep punching it with a stick until most of the wax is melted and floats to the surface. The bag will retain much of the extraneous matter, and you can dip the wax off the top of the water as it accumulates there and empty it into another container. This results in cleaner wax than is otherwise obtainable from old combs.

Some beemen prefer not to boil their wax at all but to keep the water in which it is being heated just under 185°, as it might become discolored at a higher temperature; and if you are handling only a small amount, as will probably be the case, there is no reason to boil it, even at a slow simmer—although in some altitudes it may simmer at 185°.

If you wish, you may allow the wax from marred combs or combs that have to be discarded because they contain drone cells (or for some other reason) to accumulate, and render it when you have enough to make it worth while. It should be quite free from honey, of course, lest it attract bees and ants. Remember that the bees can clean it up for you.

Figure 10

Solar Wax Melter

When your bees happen to be secreting a lot of wax early in the season, you may take advantage of this by adding a super of frames next to the brood nest. Here, temporary combs may be constructed—and this is one way to discourage the swarming instinct. You may put small bits of starters in these frames, using up broken and marred pieces of comb foundation. After the bees have completed these and drawn them out into cells, but before they start storing honey in any degree, you may cut the comb out and add it to your wax store. Leave a strip about an inch wide along top and sides; then you may use these already prepared frames for the same operation another time. You not only discourage swarming by doing this, but you get an extra pound of nice wax without using any honey except what the bees consume while building.

BLEACHING WAX. In commerce there are various methods of bleaching wax, but if you live where there is a lot of sunshine, that will suffice. Perhaps you would even like to experiment with a solar wax melter. It's fun, but so far nobody has made one which is 100 per cent efficient in getting all the wax out of the combs. The sun extractor or melter is just a black box, tilted to follow the angle of the sun, and fitted with a tight, double-walled glass top. There is a tray to put the combs in (enamelware, stainless steel, or aluminum), fitted with a lip, and a receptacle for the wax as it melts. Perhaps you could design one yourself better than any so far made. (Figure 10). A Sun Extractor has sometimes been used to extract honey as well as wax but the operator must be very careful lest it become too hot and melt wax along with honey.

If you have kept your brood nest clear and followed all the other directions given here, and *if* the nectar flow and the weather have been good, you will have a honey crop to be proud of, even with only one hive.

CHAPTER SIX

Wintering Your Bees

1. Man Is the Bees' Worst Enemy

SOME casual beekeepers—who keep bees as a sideline and care little about method—treat their bees very badly by not wintering them properly. If a colony has an adequate supply of food, it should not be tampered with late in the season. When the bees are permitted to fill the brood chamber with a good store of pollen and honey, they will arrange the food in such a manner that it will be available during the late fall and the winter and spring months. Unless this has been done, it is best to place on top of the brood chamber a shallow food chamber well filled with sealed honey. Do not be so anxious to secure a large crop of honey that you do not leave the bees enough to winter on. If you are in a cold climate where no nectar at all is gathered during the winter months, I would advise you to leave them twenty to thirty pounds—six to ten frames—to carry them through the cold weather, the amount depending on the size of your colony. Be generous your first year; experience will teach you how your bees get along during your local winters. If you rob the bees of winter stores, feeding is too likely to be neglected. It is much better to give additional feed while the weather is still warm enough for the bees to store it where they want it, and ripen it before cold weather.

When spring comes, it will require approximately one frame of honey to produce one frame of brood. There should never be less than fifteen pounds, approximately five framefuls, in the hive at any time. A smaller amount will discourage brood rearing.

2. *Are Bees Cold-blooded?*

Bees are commonly called "cold-blooded" animals, but there is a question whether this designation is correct. I prefer to think of them as warm-blooded. They are certainly not cold-blooded in the sense that toads and snakes are; and they do not hibernate. Beekeepers have worked out the matter of hive temperature in great detail. The intense summer activity of the worker bee—the exciting operation of its bodily metabolism from its great consumption of energy food—wears out its body quickly and brings it soon to death. However, during the winter, the bees remain clustered in the hive, eating very little as long as weather conditions are good. A whole colony of bees will consume only a few ounces of honey per month when loosely clustered at the optimum winter temperature of 57° F., in contrast to the large amount consumed during the busy breeding season, when the warmth within the hive must be raised to 95°—the optimum brood-rearing temperature.

So, whether or not the individual bee is cold- or warm-blooded, the colony is definitely warm-blooded. It can and does change the temperature within the hive according to its activities.

3. Preparing Your Hive for the Winter

When the honey flow is over and you have harvested your crop; when the days grow shorter and colder and the nights colder and longer; and when you see the worker bees trying to expel from the hive the few remaining drones—then is the time to think about how you will prepare your bees for maintaining themselves over the winter and for coming into spring in such good condition that they will build up to prime honey-storing strength before the nectar flow. Besides seeing that your colony has enough food, you must take care that the hive is in good condition and well ventilated to avoid excessive moisture

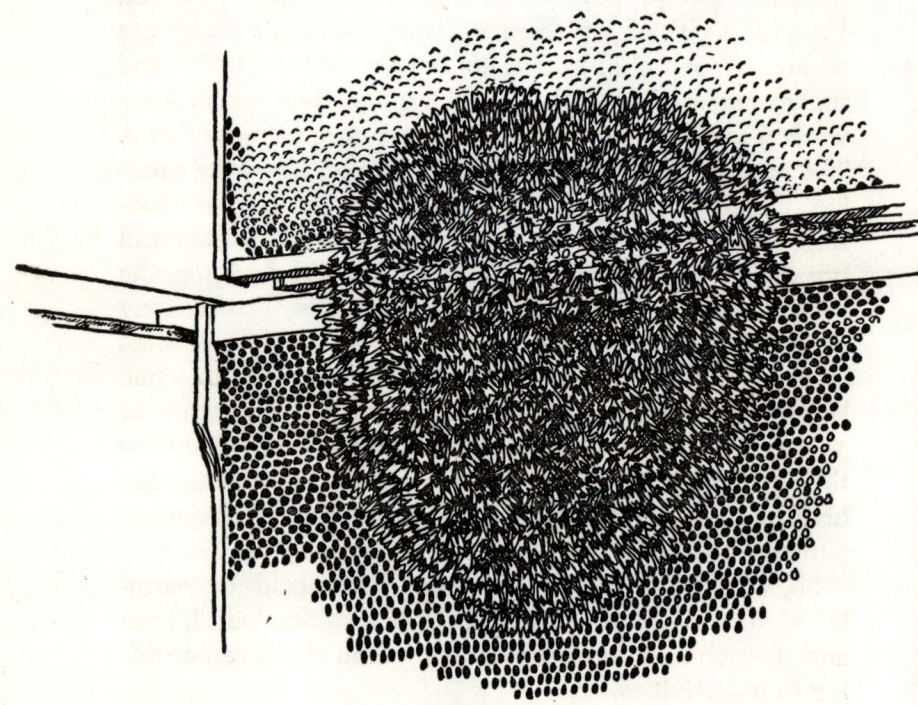

Figure 11

Clustering Bees

accumulation—and that the colony is large enough. A strong colony will form a larger cluster that will present greater resistance to cold than a small colony can furnish. Moreover, it will consume less food per bee than will the small one.

Because local conditions will have a great deal to do with the manner in which you prepare your bees for the winter, it will be advisable to consult some long-time beekeeper in your neighborhood. If you live in southern California or in one of the Southern states, no special preparation for wintering will be necessary beyond seeing that the bees have enough honey. If they can occasionally gather a bit of nectar—as from eucalyptus trees—during the winter, fifteen pounds may be sufficient to carry them through. When the California radio announces the necessity of smudge pots for the orange ranches in your neighborhood, you might cover your hive with a blanket or some old newspapers for the night. It is not really necessary, but if you do, be sure to uncover it, when morning comes with its California sunshine. A commercial beekeeper could not take the time to do this for all his colonies!

However, if you are in a climate where the winter temperature ranges from below zero to above freezing, and where it may be below 20° F. for weeks at a time, it will be a good idea to insulate your hive if you are going to leave it out of doors—unless you have one of the double-walled hives or an aluminum hive, both of which are cooler in summer and warmer in winter. Even these might need insulation in some localities. If the humidity is low in your territory, it may be enough to wrap your hive in black building paper; remove the top cover for the wrapping and replace it after the hive is wrapped. There should also be upper and lower entrances, for ventilation; and the bees should have plenty of stores. If you find that the combs on the ends of the brood chamber are full and that there is a good supply of honey in the central combs, in the tops, and in the corners, you can feel reasonably confident that the hive has enough to winter on. If your colony is a strong one, as it should be if you have helped it to maintain a clear brood nest, it will cover from six to eight frames, and the center of the brood nest should be filled with eggs, larvae, and sealed brood. There will also be cells filled with pollen, and the bees will be bringing in more all the time until the cold weather sets in.

If your beekeeper neighbor has advised you to cover your hive with a "quilt," and you have done so, be very careful to see that it is porous enough to allow the escape of damp air. Bees are practically defenseless against excessive humidity—they can endure much more cold when they are dry. Fresh air and good ventilation are therefore very important. The "quilt" may be made of ticking, old carpeting, cotton twill, tent cloth, or anything of that nature. If the carpeting is wool, however, do not put it next to the hive, but on the outside of the rest of the packing, as it might attract wax moths which would make headway into the hive before you observed them. They don't object to a bit of wool with their wax!

4. *Upper Entrances*

There are various ways of furnishing an upper entrance for ventilation and the release of excess humidity. The super cover has an oval hole in the center to receive a bee escape. If this is used on top of a hive of more than one story, a portion of the front rail on one end of the cover may be cut away to provide an upper entrance. When the hive is wrapped with building paper, a slit may be cut in the paper to allow exit from the bottom entrance or from this hole; or the bees may be left to find their way out, when they want a cleansing flight on a warm day, through the top folds of the paper. In any event, ventilation would be provided through these folds.

The part of the front rail which is cut away may be used to make a gate which can be swung on a nail pivot to enlarge the opening or to close it as required. (See Fig. 12.)

Or the cover may be turned upside down, with the gate swung backward. This keeps the air from flowing too rapidly over the frames of bees. Also it gives room enough —really a double bee space—for the feeding of pollen sub-

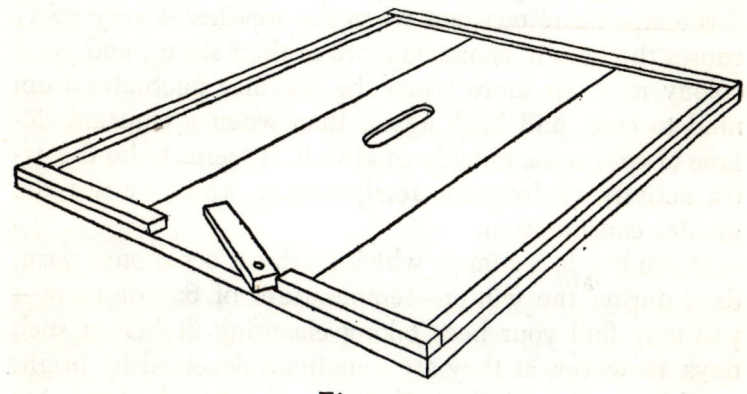

Figure 12

Gate in Front of Super Cover

stitute if necessary. It allows the warm air to escape and carry with it the excessive moisture from which we wish to protect the bees.

In lieu of such an entrance, the cover may be lifted up slightly, so as to leave a crack at the back to allow a current of air to circulate all through the hive. Also, a super or a cover may be slightly staggered, but this is done more commonly to provide ventilation in very hot weather rather than in the winter.

5. Winter Temperature of the Hive

As winter approaches, the queen lays fewer and fewer eggs and the flying bees do not venture far from the hive, and then only during the warmest hours of the day. After the first frosts, the bees are usually seen to make only brief flights, to void accumulated feces and maintain the cleanliness of the hive. They will begin to cluster when the temperature falls to 57° or below, and the cluster will contract or expand as necessary to maintain this degree. The necessity of greater activity (to keep this temperature constant when the weather is very cold) causes the bees to consume more of their stores; and yet a colony may eat more when the weather fluctuates from mild to cold, and back again, than when a constant degree of cold is the outside condition. It seems to be the extra activity of frequent readjustment which causes the greater consumption.

If you live in a climate which produces occasional warm days during the winter—temperatures of 60° or more—you may find your bees taking cleansing flights on such days. However, as they are sometimes deceived by bright sunshine at the entrance, they may fly out when it is too cold. So the entrances should be darkened and kept so during most winters.

If you winter in a cellar or in a shed where an outside temperature of about 45° can be held, it will be unnecessary to insulate your hive. Do not winter in a basement containing a furnace, as that would be too warm. The temperature should not go above 50° or below 40°, and the place should be kept dark. One hobbyist I knew, who did not care how much money he spent on his bees, used to winter three hives in an air-conditioned, dark room; but once a month during his cold winters he would turn on a light (first having heated the room to 60°) and entice his bees out for a cleansing flight. Apparently he did not mind cleaning up after them! He felt that they reached the spring season in better condition because of the "coddling," but of course this would not be practicable for the commercial beekeeper.

6. *Feeding*

Some beekeepers feed their bees during the winter a sugar syrup or a special kind of sugar candy. This may be necessary to stimulate brood rearing in early spring, when there is no natural source of honey at hand, and when winter stores are all consumed—or when the weather is too bad for the bees to go out and gather nectar from blooming orchards. A hundred pounds of honey might not be too much to carry a strong colony through a severe winter followed by a blustery, wet spring, and in such a case feeding would become necessary. Dry sugar, sprinkled on a sheet of paper above the frames, may be used and is less likely to cause robbing than sugar syrup. Candy made of pure granulated sugar, unflavored, may also be placed above the frames, or lump sugar if there is no starch in it; but any dry food requires moisture within the hive, or the bees will be unable to avail themselves of it. If you have no fear of robbing, you may make a sugar syrup of one part sugar to two of warm water, by bulk. The warm water helps the sugar to dissolve, and if very warm must be cooled before being given to the bees.

If you have enough honey left in your hive, you will not need to feed in fall or winter; but it might become necessary in the spring. There are several types of feeders on the market. However, you may use an ordinary small pan, placed in the upper story between the frames and the reversed cover and filled with syrup. Put a strip of dampened cheesecloth over it, to enable the bees to take the syrup without becoming submerged. If there is any danger of robbing, be especially careful not to spill syrup near your hive.

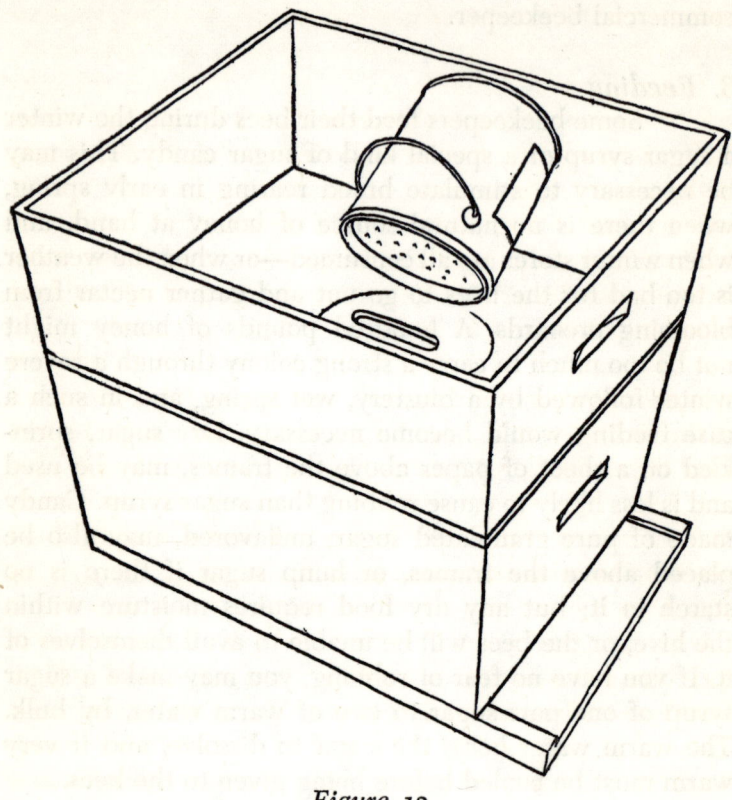

Figure 13

Feeding the Bees

Some bees will die during the winter, and the survivors cannot always carry out the bodies. It is wise to see that the entrance is kept clear of dead bees, so that ventilation will not be obstructed. A small slender stick or a piece of wire may be used to sweep them away.

7. "Bee Weather"

Beekeepers recognize that, if the bees do not find the temperature of the hive to their liking, they will produce "bee weather" to the best of their ability. When the colony is most nearly dormant in winter, the temperature within the loosely clustered mass is at least 57°. Therefore, in wintering bees in sheds or cellars, if you keep the outside temperature carefully controlled at 45°, the extra twelve degrees of heat will be added by the slight activity of the cluster even when least exerting itself. When bees are wintered out of doors in cold climates, the insulating packing keeps the heat manufactured by the bees from escaping too rapidly, whether the packing is of building paper or some kind of quilt.

As soon as the inside temperature drops so low that the optimum 57° is not maintained by the natural metabolism of the clustered bees, the bees increase their activity and their consumption of honey. They convert the honey into heat by moving wings and muscles and by stirring about. As long as they have good honey within the cluster, they can keep themselves warm even though they do shorten their lives by wearing out their bodies.

One exception should be made to this. If the honey is very dense or jellylike or candied very hard and the bees have no water to dilute it, they may die of starvation even when clustered on a lot of honey. I have saved such starving, freezing colonies by feeding very thin, warm syrup. Usually the bees liberate enough water from the honey they consume to liquefy and dilute other honey; but there are exceptions, and I think this need for water is not so generally thought of as it should be. I have seen the results of water lack in the colder areas of the mountain states, and they might occur in other places. In such instances, colonies may be found dying when there is a lot of unusable honey within the cluster, and their owner may think they are freezing to death. Feeding thin syrup or even sprinkling the combs with warm water soon brings the surviving bees to life.

8. The "Warm Way" Entrance

Some hives are constructed with the entrance running parallel to the frames, instead of at right angles to them. This is a common practice in England, I understand, although not so frequently seen in the U. S. Some years ago, the bee journals of America gave a good deal of space to discussing the topic, as it was claimed that such an entrance was better for wintering—then no more was heard about it.

I have not had an opportunity to test the theory out in a bitterly cold climate; but there is another matter to be considered which I have had called to my attention again and again in northern California coastal regions, where winds blow chill and cold; even in August the minimum night temperature is sometimes as low as 40°, while the maximum day temperature is frequently as low as 60°. This is not good bee weather. With the usual hive entrance, the wind blows in viciously between the combs and the bees draw up in a cluster every night at the top of the hive, away from the cold, unless furnished with supplemental heat or given the "warm way" entrance.

Moreover, we spend time and money trying to get perfect combs drawn out on wired foundation in good weather; and when we try to keep a clear brood nest, we probably get these fine worker combs filled with brood to the top bar and to the corners just once. Then comes a spell of cold nights and cool days when the bees do not build comb and the queen lays no more eggs in the corners of the ten frames facing the "cold way" entrance. What happens then? The comb is cut away in bits by the bees to use elsewhere, and a triangular hole is left in the corner. It is ruined forever as a perfect brood comb.

If we reverse the comb to get the hole filled, we may succeed, but it will be filled with drone comb in which drone brood will be reared. Worse than that, the corner we have moved next to the cold entrance is also cut out, and we have two ruined corners in the comb.

Finding this situation recently, I changed some hives to a bottom board having the entrance on the long side and found this to be a great improvement for my coastal section of northern California. Only one frame, that next the entrance, can be damaged. No corners were cut out after I made the change, and I had nine perfect worker combs.

I made new floor boards of ¾-inch waterproof marine plywood, and this obviated the warping which has heretofore been held up as an objection to the use of the entrance on the long side of the hive. The flooring may also be made of aluminum. The grooved cleat into which the floor board fits is admittedly frail when made of ⅜-inch material, but there is no good reason that 1⅛-inch strips should not be used. A heavier cleat, used either with waterproof plywood or with aluminum, makes an excellent bottom board.

It would be easy enough for a manufacturer to make the kind of bottom boards which would free us from trouble. With either marine plywood or aluminum, it would be simple to have two short cleats grooved so that the floor would slant downward from about ¼ inch below the frames at the rear to ½ or ¾ inch at the front; this would give a slope to the bottom board without tilting the hive. Then the hive may be set level, and the frames will hang perpendicular, as they should.

CAUCASIAN BEES WITH THE COLD WAY ENTRANCE. Also more noticeable as the use of Caucasians is increased (and I have some of them) is that, when the entrance is at the end of the frames, all ten combs are so firmly fastened to the bottom and ends of the hive at the entrance that they can be taken out only with considerable effort. Sometimes the bottom bar will be pulled off the frames, much to the beekeeper's annoyance. This is bad enough with Italians, but with Caucasians the trouble is increased, especially if the hives are somewhat worn and the entrances higher than ⅜ inch. I have scraped from the ten frames of a Caucasian colony as much as one pound of the propolis mixture used in fastening the frames down and closing up the entrance in preparation for winter. This occurs where winds are bad and the entrance is at the end of the hive. I have found that the bees use much less propolis if the entrance runs along the side of the hive and is partially closed for them by the beekeeper when winter comes.

I do not think that anyone who has ever used entrances parallel with the combs will ever willingly go back to using end entrances.

9. *Supplemental Heat*

When winter temperature continues extremely low over a long period, the colony might be doomed unless furnished with supplemental heat. The bees cannot break the cluster to move to another part of the hive where honey is available—they would freeze if they separated—so they starve and freeze eventually, anyway. Bees can combat a temperature of 100° below zero if they have food, but not for long. To do so, a colony must be very strong and consume pounds instead of ounces of honey. It must exert a destructive metabolism. The bees seem unable to reason that, at some point, they should break up the cluster and move together a few inches to another honey supply; rather, depending upon instinct, they die—whether of cold or of starvation.

The only thing that will save a colony exposed to great cold over too long a period is supplemental heat. With good food and warmth, a colony can be saved. Whether or not it is profitable to supply bees with additional heat in cold weather depends on the producer, his method of management, his location, and the availability and cost of heat.

10. Equipment for Warming the Hive

If you are really serious about wanting to use supplemental heat, you should know why and how, and for full information, I suggest, not too modestly, that you secure the second edition of my small book entitled *Scientific Beekeeping*, prepared in collaboration with Dana F. McFarland. This will give you the scientific data you should have for an adequate understanding of the subject. However, the practical information which you need is given right here.

One of the first things to note is that we want to heat the hive, and not the bees. Hence we need a large surface area which can be kept at a moderate temperature, an area which will emit long wave lengths of heat energy. The source of heat, therefore, should not be in the hive with the bees, but should be in a mixing place where the air is heated to the desired degree and then introduced into the hive. The honeybee cannot function under direct radiant heat. If the cluster were exposed to radiant heat during the winter, the outer layer of bees would be warmed while the air in the hive, as a whole, would remain cold. So it is necessary to supply the bees with heat as it is given to them by nature—by warming the air before it passes into the hive.

116 | WINTERING YOUR BEES

So a heating unit is required which is located not within the hive, where the infra-red radiant-heat energy would directly strike the bees, but in a chamber beneath the hive. If you cannot buy such a hive-heating unit, you can make one. Prepare a shallow box or drawer, about 4 inches deep and 14 inches square. See Fig. 14. This will be suitable for any hive of standard 10-frame width or more. Inside this

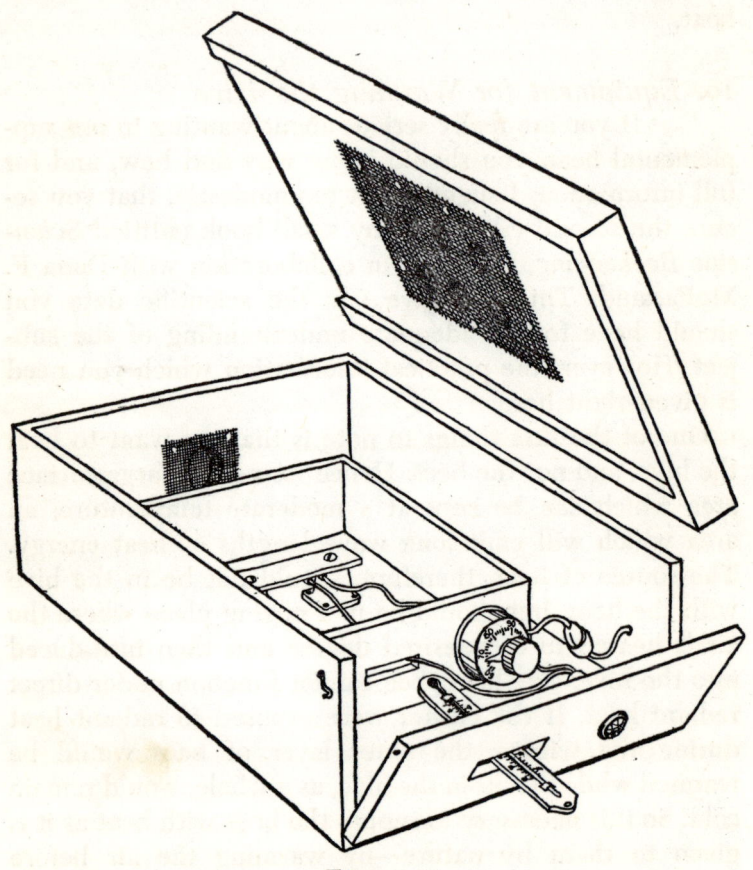

Figure 14
One of Several Possible Hive-Heating Devices

EQUIPMENT FOR WARMING THE HIVE

drawer you will want an electric heating element with a thermostat and attachments. The thermostat should be controllable at settings of 45° and 85°—the former for ideal winter conditions, and the latter for brood rearing.

Place the drawer within a shallow super or other box, just the size of the hive, and five to six inches deep. The box should have a bottom to keep out pests, but no top. It serves as a temperature-equalizing chamber through which fresh air passes and is warmed before it rises naturally into the hive above. The thermostat, properly set, will insure that just as much warm air rises as the bees need.

The rear end of this equalizing chamber has a hinged door which may be opened for access to the heating unit. Through this the drawer containing the heating unit may be inserted or removed at will, without disturbing the colony. It also allows adjustment of the thermostat without removing the unit.

In the hinged door at the rear of the equalizing chamber, bore two holes; one for the pilot light—which will tell you at night whether the heat is on or not—and the other, directly in line with the same kind of hole in the unit drawer, for the thermometer, which may then be inserted from the outside for checking purposes.

After the box or shallow super has been prepared, set on top of it a bottom board with holes cut in it and screened (as shown in Fig. 14). Then, removing the original bottom board, place the hive containing the bees on this especially prepared one. When a hive is on this unit, the entrance may be open or closed as circumstances require. In fact, from then on, the colony is handled in conformity with your own system of management.

At the front end of the equalizing chamber (which is also the front of the hive), you will have a 1½-inch hole (screened to keep out vermin) for the passage of fresh air. Cold air enters this screened hole, is warmed in the equalizing chamber, and passes the heating unit, the heated walls, and the warmed floor board of the hive above the equalizing chamber. Then, as it is needed, the warmed air rises through the screened holes you have made in the bottom board and circulates through the colony at the desired temperature. As the air within the hive cools, it falls to the bottom board, passes out through the entrance, and carries with it all the noxious gases from the metabolism of the bees, to be replaced by fresh warm air from the equalizing chamber.

11. *Setting the Thermostat*

The thermostat may be adjusted approximately before putting the unit in its place in the equalizing chamber, and the final adjustment may be made after testing with a thermometer in position. For spring build-up, set it at 85°, as the bees themselves will bring it up to brood-rearing temperature by their metabolism. This point of 85° may be permitted to vary a little; but when adjusted at 45° for wintering, the thermostat should be set very accurately and made lower and lower until exactly the right point is reached, if possible.

When a colony is first placed on such a unit as this, the thermometer may appear to act erratically at first and may not steady for a day or two. The bottom board, indeed, the whole unit, must become heated to the desired degree, and the bees must become accustomed to this uniform degree of heat energy provided for them before they can make full use of it. It will require a few days, perhaps, for them to realize that this is a warmth that may be depended upon.

Another point: the circulation of air within the hive has to take a new course. Formerly the entrance served both for the admittance of fresh air and for the exit of used air, but now, apart from allowing the bees to come and go, it will be used only for ventilation and for an air outlet. So do not be disturbed by temporary variations as revealed at first by the thermometer in the equalizing chamber. Soon all will become equalized.

12. What Supplemental Heat Does for You and the Bees

Many beekeepers move their colonies in the spring to locations where a minor early nectar flow will cause their bees to build up to honey-storing strength. Or the bees might have to be fed to encourage brood rearing, and even then they might not reach the right strength at the right time because of cold weather. Then they may be too weak for good results in pollination when the orchards are in bloom.

However, when supplemental heat is used, this extra work and these disadvantages need no longer trouble the beekeeper.

Moreover, supplemental heat makes the double-walled hive unnecessary. In some cold regions there are good reasons for giving the bees the extra protection of double walls; but there are also disadvantages: by the use of double walls or heavy packing, conditions are created like those in the interior of a refrigerator; and the sun never shines long enough during the cold days to warm up such a hive and allow the bees to have a cleansing flight, as they would have from a single-walled or unpacked hive. They might not even be able to break the cluster and move a few inches to where there is plenty of honey, and thus they might starve in the midst of plenty.

By the use of supplemental heat, however, the advantages both of packing and of single-walled hives may be retained. The beekeeper may pack his hives as heavily as he wishes and then, by thermostatic control, keep the bees at the normal winter temperature of 45° (which they will raise to 57°) until, on some fine day in winter, when the sun warms the outside air to about 60°, his bees will fly as freely as would the usual colony in a single-walled hive.

With differential thermostats, the procedure is very simple. One thermostat, set at 45°, is used to control the temperature inside the hive; and this is combined with another outside the hive which, as the day becomes warmer, gradually makes the hive warmer inside until it reaches the 60° or more which obtains on the outside. Then the bees fly as freely as they would from a single-walled hive warmed by the sun.

Each locality presents its own problems, and from the suggestions given here you should be able to select the method of wintering which will be best for your colony. You will prove yourself a good friend to your bees, and they will reward you by coming into the next spring in good strength for the work of the season.

CHAPTER SEVEN

The Next Spring and Thereafter

1. Package Bees

The following season, having had some practical experience in beekeeping, and having wintered your colony successfully, you may order some package bees and build up another colony or two, if you wish. Package bees are ordered by mail during the winter, with shipping date stipulated, and are usually shipped by express, although a small shipment may be made by mail. You may also buy one or two more hives at the same time, with supers and frames, all in the flat, and assemble the lot yourself, following the accompanying directions. This will give you opportunity to observe many features of hive construction which you might have missed before.

Your new bees will enable you to compare their behavior during the season with that of your wintered-over colony.

A package of bees is practically equivalent to a swarm. The queen will not be laying at her maximum rate, and under "natural" conditions such a colony would find a home in an empty box or in a hollow log or tree, without combs. The colony would be in balance, because the queen would keep up with the comb building, filling it with eggs as it was built.

So, if you put a package of bees on a fully drawn-out set of combs, with honey coming in or with plenty of food, you will throw it out of balance, because the queen is not laying actively enough for such conditions. Then the bees might behave as if the queen were failing and take steps to supersede her.

So the best plan is to get your package early in the season, before there is any flow of nectar to speak of. Then put in some *full combs of honey*—three or four—and leave them alone until the main honey flow is on. In this way the bees will have little space for brood rearing except as they use up the honey and make room. So the queen will step up her egg-laying gradually, and the colony will remain in balance.

Colony balance is very important in avoiding supersedure, and this is often a critical problem with package bees.

The package is a cage of wood and wire and contains from two to six pounds of bees, their queen suspended in a mailing cage. The larger packages are preferred when the bees are wanted mainly for pollination. The bees are supplied with a can of sugar syrup for food during the journey.

When you are expecting a package at the express office, warn the agent not to leave it in the hot sun. It is well to inquire daily about its arrival and not leave it to him to notify you. Examine the package for dead bees when you receive it. If there are many, a report to that effect should be made on the express receipt, to facilitate damage claims. Accept the shipment anyway; the shipper will make the loss good if it is serious, and to refuse it would complicate matters.

If your package arrives too late to be hived the same day, put it in a cool, dark place until the next day. Late afternoon is usually the best time for hiving, as the bees will not tend to fly so much and will become settled overnight.

If you are putting the package in a hive with three or four frames of honey, as I have advised, it is best to release the queen at once. Close the entrance to the hive with a strip of lath for a while. Remove the feeder can from the package, and take out the queen cage also. Open the latter so that the queen can find her way out easily, and place it on the bottom board under the full frames of comb, screen side up. Then put the package—open side down—in the empty space beside the frames, on supports at least as high as a couple of bee spaces, and close the hive. It will not be long before both queen and bees will busy themselves inspecting the combs of their new home. You may open up the entrance a little, after a few minutes. If a few bees fly out, it will not matter.

The next day you may remove the empty package.

If you are unable to supply your new bees with three or four full combs of honey and pollen and must rely on frames containing foundation with little or no honey, a different method of putting the bees in the hive is suggested. This operation may be begun earlier in the day, in good weather if possible. First reduce the size of the entrance to just about enough to admit two bees at a time. The hive should be where it is to remain. Take the feeder can and queen cage from the package, and place the queen cage—the paper over the candy end removed, and this end up—between two frames. Spray or brush a thin sugar syrup over the wire of the cage or sprinkle all the bees with warm water. The cage may be bumped on the ground to break up the cluster and make it possible to wet them all. This seems to quiet them down and make them easier to handle, as it gives them a job to do in cleaning themselves off.

Then place the opened package near the hive entrance. The bees seem to know that their queen is in the hive and will gradually enter and orient themselves, showing little tendency to fly before becoming used to their new home. Some of the bees will start to eat out the candy, and before long the queen is released.

When the package is introduced in this way, a ten-pound feeder can full of sugar syrup should be placed in a super above the frames. An ordinary can, holes punched in the top and punched side down, is sufficient. If there is some nectar and pollen to be had in the fields, this will be all that is necessary to get the colony started; but if this becomes exhausted, or if your full combs of honey become exhausted, continue feeding until the bees are able to gather what they need.

Old bees tend to acquire dysentery and even young bees may acquire nosema when fed sugar syrup, so it is well to forestall this eventuality by adding "Fumidil-B" to the solution. This is the trade name for the soluble salt of an antibiotic which has been found to possess a high specific activity against *nosema apis*. It prevents the reproductive stages of the parasite from attacking the epithelial cells of the bees' stomachs. It has no effect on the spores and so must be given in the food for three or four weeks. It is supplied in bottles containing just enough for a ten-pound pail of syrup. Heat the right amount of water almost to boiling and add the powder. Stir until dissolved and add the sugar and stir until it becomes clear.

There are various modifications of these two methods, and you may eventually adopt one which suits you better than either.

If there is no pollen in the combs of honey you have given to your package bees, and none available in the fields at the time of introduction, it may be necessary to furnish them with one of the pollen substitutes. There are several formulas for these. A comparatively simple one which you may mix up yourself is made of six tablespoonfuls of soybean flour (expeller processed) and one tablespoon of brewer's yeast, animal type, moistened with honey or with sugar syrup, half water and half sugar. Knead until it is of the right consistency to press into a thin cake, and place the cake over the top bars above the brood nest. After that, the package colony should be handled as little as possible until brood rearing is well advanced.

If you live in a locality where there is a surplus of pollen at certain times of the year, a pollen trap at the entrance to the hive will give you enough pollen to meet emergencies; you can save the pollen until needed, feeding it straight or mixed with the pollen substitute.

About twenty-one days after hiving your package, the adult bees will reach a low level and new brood will be at a comparatively high level. There is danger of supersedure at this juncture; but if you can give this new colony a good comb of emerging brood and bees from your other colony, placing it next to the brood in the package colony about two weeks after hiving the package, this danger will be obviated.

2. *Your Wintered-over Colony*

If you have wrapped your other colony in building paper for the winter, or otherwise insulated it, the insulation may be removed in early spring, but not before the daily maximum temperature is about 60°. If your colony has been heavily packed, it might have expanded its brood over several frames, and unpacking too soon, unless you have supplementary heat, will cause a loss of brood from chilling. With supplementary heat, you may unpack at any time and just leave the heating unit connected.

We have spoken before (page 76) of "spring dwindling." When a colony has not had enough stores for the winter, the loss of adult bees is not compensated for by brood; but when there has been enough food, the loss of old bees will soon be made up by the rearing of brood in late winter and early spring in such proportions that the old ones are soon replaced. If, however, dwindling becomes very evident during the spring, the old queen should be removed and replaced from a nucleus or by a queen in a mailing cage. With two or three colonies, you may use one to help another according to directions previously given.

Sometimes a colony can be brought up to strength very quickly by giving it one or two additional combs of brood and bees from another colony.

If your wintered-over hive is of two or more stories, you might find the brood in the upper body at the end of winter; and the queen might keep on working there. If you have two full-sized hive bodies, or one body and one super, you may reverse them when the upper body is filled with brood comb. This will make the queen move up, which she is more likely to do: it seems to be the nature of bees to go up rather than down!

The bodies may be reversed again when the top story is filled once more, and supers may then be added.

If you happen to have drawn a queen who confines her egg laying to the center combs, leaving the side combs untouched—something which might happen, especially in colder weather, without supplementary heat—you can work the side combs into the brood area, if there is no real danger of chilling the outer brood. Put an empty comb next to one filled with brood, and place the storehouse comb beyond that, toward the outside edge.

When the orchards are over their blooming period, but before the main honey flow in a locality starts, there may be danger of the hives' becoming depleted of honey and pollen. They should be examined for a possible need of feeding at this time.

3. *Spring Build-up*

In the spring, the bees need a plentiful supply of pollen to begin brood rearing. If, on a bright spring day, you see the workers coming in with the pollen sacs on their legs well filled, you can conclude that brood rearing will be well under way in a few weeks. Do not open your hives until it is well advanced—some three weeks after you observe the loads of pollen. Then give them a careful inspection. If you find several combs partly filled with brood, all is well. If not, something has happened to the queen, and the colony will be weak. If you have two colonies, the best plan is to unite them, whether one is weak and one strong or both are weak. You may place the bees and combs of the queenless colony in a super, and on top of the good colony spread a newspaper, punching a few holes in it. The bees in the super will gradually mingle with those in the main hive. Be sure that neither colony has any disease before uniting them.

Your united colony should have six or seven frames of honey in the food chamber, if possible, unless there is a light honey flow on. Then four frames will be enough. If you or the bees cannot furnish enough honey, feed sugar syrup.

In localities were there are many cold, blustery days in the spring, and where the weather may be cold and wet even when the orchards are in bloom, the use of supplementary heat is very helpful, as I have explained in chapter 6.

4. Conversation Pieces

I want to offer now some odd items which are good to know and to talk about with other beekeepers and with your friends. Some of these items repeat—with different emphasis, perhaps—what has been said before in this book, but they are worth repeating! Others are quite new.

1. BEEKEEPERS' ASSOCIATIONS. If there is a beekeepers' association in your county or state, visit the meetings if you can and do not be afraid to ask questions. Beekeepers are always ready to talk bees, even with the amateur. In my long experience, I have found less professional jealously among beekeepers than in any other field: they seem always ready to help each other.

In addition to the county and state associations, there are a large number of regional groups. The American Beekeeping Federation helps its members in obtaining their legal rights, and aids in securing legislation for the betterment of the industry. The names of officers and headquarters of the nearest associations will be sent to you upon request. Write to the Agricultural Research Service, Section of Beekeeping and Insect Pathology, Beltsville, Maryland.

As an amateur, you might not feel called upon to join a national group, although you would be welcome; but your local association might be very interesting and helpful to you.

2. YOUR COUNTY INSPECTOR. Also become acquainted with your county bee inspector. He should be notified of your ownership of your first colony, and its location. He will always be glad to help you with your problems.

3. USE OF CHEMICALS. I suggest caution in the use of chemicals for the cure or prevention of bee diseases. It may be, as some allege, that the burning of equipment exposed to AFB will eventually be outmoded as more is

learned about the use of the newer drugs, and as beekeepers learn from experience what results to anticipate. But until this is certain, burning will remain the safest method. The sulfa drugs have been proved better as preventives than as cures, so far; and they have even been alleged to spread AFB when a colony was already infested. So—rely rather on good practices and regular inspection.

The beekeeping industry is now producing improved resistant strains of bees through selective breeding, and the destructive diseases may eventually be eliminated.

4. THE NUMBER OF BEES PER COLONY. There are about 130 square inches in a Langstroth frame, or 260 square inches in a frame full of brood on both sides. It is estimated that 50 bees will emerge from each square inch of worker comb, 25 on each side, in ten to twelve days after the brood cells are capped. So if you make an estimate of the number of capped cells in a hive on a certain visit, and again at your next visit ten days later, you can figure on the number of bees that will emerge during the following ten days.

Of course, the number of workers per colony will vary considerably with season and weather. A normal colony in spring should have 15,000 to 25,000 bees. The number should increase to 50,000 or 60,000 by June. In the fall, the number will gradually diminish as colder weather approaches. In a honey-producing colony, there should be at least 15 pounds of bees for the best results—upwards of 60,000. As it requires 20,000 to 25,000 young bees for the housework, the more bees there are, generally speaking, in excess of the housekeepers, the more honey you will get.

5. A LATE SWARM, or a new colony established late in the season is unlikely, if left to its own devices, to be ready to store honey until the main honey flow is nearly at an

end. Therefore, such a colony would starve the following winter because it would not have gathered enough stores to live on. That is the basis for the old saying:

> A swarm of bees in May is worth a load of hay;
> A swarm of bees in June is worth a silver spoon;
> A swarm of bees in July is not worth a fly.

However, the modern beekeeper can take that July swarm and easily make it into a colony that will go through the winter and store honey the following season; or, if there is a honey flow late in the season, he might also get a crop for himself. If the worst happens, he will not let the swarm die of starvation but will feed it enough sugar to carry it through the winter.

6. SYRUP FEEDING. The ordinary friction-top pail is as good a syrup feeder as any. A few nail holes punched in the lid will enable the bees to remove the syrup. Avoid leaky lids. If the syrup escapes too rapidly for the bees to take, it will be wasted or will start other bees to robbing. Turn the pail upside down over the hole in the inner cover, put an empty super over it, and place the hive cover over all.

7. GOOD COMBS ESSENTIAL. Good combs represent the greatest cost in equipping your colony, but they are a profitable long-time investment. When you have good all-worker combs, you will not need as many to keep the colony going, which means greater profit in the end. If you have drone comb, you will have fewer cells in a given space. A hundred drone cells occupy the same space that 150 worker cells would occupy. It is therefore a good thing to cull your old, broken, discolored combs after extracting and use them for beeswax.

8. HOW MANY BEES CAN ONE MAN KEEP? In most localities, one man can keep 500 colonies by himself, as a full-time job, with a little extra help in the busy extracting

season. As a sideline, one man can handle 100 colonies quite well, with extra help occasionally.

My old friend George Bohne, one of my favorite beekeepers, used to carry an assortment of colored thumbtacks on a pad in his pocket. He used them to mark good and bad combs. Perfect brood combs rated a white thumbtack; those not quite so good a blue one beside the white one, while a red one was placed on any frame not good enough to be used in the brood nest. This frame was transferred to a super and finally, when empty, to the melting pot. He refused to be tempted, no matter how full of stores it might be, ever to put a comb with a red tack into the brood nest. I believe this was the foundation of his success in keeping bees, although he never had more than 100 colonies of his own or an investment of more than $3,000 at any time. He loved beekeeping and would never keep enough bees to make the work burdensome. He always had another occupation, and bees were a sideline. During twenty years he sold $100,000 worth of honey, $80,000 going to one firm, according to their books.

9. How much honey can a colony make? Instances have been known of more than 500 pounds of honey being stored in one season by a single colony. This is not common—or every beekeeper would be a rich man. In the coldest climates, the colony will consume, on the average, 50 to 60 pounds during the winter. The remaining 50 to 450 pounds over and above what they consume is of absolutely no use to the colony. Yet the bees always gather all they can, to the advantage of their keeper.

The twin-colony hive will store proportionately more honey than the single-colony hive.

10. The twin-colony hive. There are various methods of developing a twin-colony hive; such a colony often produces wonderful results for automatic requeening, swarm control, and honey-crop increase. My own favorite

two-queen hive is long rather than built up and is very easy to operate. If you are interested, you may build one yourself from the specifications advertised in the Appendix (p. 141) under the title of *The Twin Colony Hive*.

11. TIME OF HIVING PACKAGE BEES. In some places package bees may be hived early in April, but in others it is best to wait until after May 1. In the Midwest, the latter part of April is usually a good time. In parts of California, they may be hived in February or March, although there may be a later cold spell which will hamper their operations even when fruit trees are in bloom. It is best to consult your local bee inspector or other beekeepers in your neighborhood.

12. TO HAVE NICE WAX. Remember that, when you are melting wax, it should remain hot for as brief a period as possible. Dip from the pan or tank as it melts, when you can. If you use a wax press, release and apply the pressure repeatedly: this does a more efficient job than constant pressure does.

13. DO BEES DAMAGE FRUIT? Beekeepers are often compelled to combat the idea that bees cause damage to fruit and other crops by sucking the nectar from the flowers. Not only is this untrue but the bees are a great aid in pollinating the flowers. This is now coming to be recognized, but there is another charge made against bees which is just as untrue. This is the charge that the bees puncture the skins of fruit—peaches and grapes, for instance—and suck the juices. Circumstantial evidence seems to point to this, as bees may be found on the pierced skin of a fruit, sucking the juice. However, if you will watch carefully, you will find that the fruit has been pierced previously by a bird or other agent, and that the bees help themselves. Bees will starve if confined in a box with perfect fruit, no matter how juicy, for they cannot, themselves, puncture the skin.

Do your best to disabuse your neighbors' minds of this often-repeated calumny against bees.

14. POLLINATION. More than fifty agricultural crops are absolutely dependent upon bees for pollination. The bees pollinate the blossoms and thus produce larger and finer fruit, grains, beans, etc. If all the bees in the world were suddenly destroyed, the population of the world would be on a starvation diet within a few months.

Pollination for seed is very important, too. I knew of a plot of 130 acres of alfalfa which had apparently outlived its usefulness and was about to be plowed out. Although the owner had never raised it for seed before, he was persuaded to try it and use bees for pollination. Accordingly, five colonies of bees per acre were moved in. A crop of honey could not be expected, but the result in seed was tremendous. Instead of the national average of 97 pounds per acre, 1120 pounds per acre were produced. With seed at a dollar a pound that was a worthwhile profit.

Many similar and quite true stories could be told of other crops.

Bees are called upon to pollinate fruit trees very early in spring, unfortunately before they have normally become populous enough to do a good job. But the use of supplemental heat will build up colonies very rapidly and will be profitable without question in many areas where pollination is the principal need. A moderate amount of heat, together with syrup and pollen feeding, will promote egg laying and provide a goodly number of young bees without causing that intense breeding which would make a colony immensely populous too soon. Too early a growth in population would require continued feeding if the main honey flow did not come as expected.

15. THE LOCATION IS VERY IMPORTANT. We have to face the fact that some places are not good honey-producing

areas. In such a situation, much of the advice I have given will not be helpful. For instance, I have stressed the need of having good strong colonies in the spring. When a colony has been wintered where there was no cold weather to confine the bees to the hives, as in orange-producing areas, they use up in profitless brood rearing almost any amount of honey left in the hive and come through the winter with many bees worn out trying to find nectar when there is none to be had. The spring season finds bees and no honey. In such areas, it is usually better to provide the bees with barely enough stores to carry them through the winter, in the hope that they will cut down on brood rearing. Then, in the spring, they may be fed when it is necessary to speed up production of brood.

The commercial beekeeper usually moves his bees from the oranges to the mountains or some cold place for the winter, if possible. This stops brood rearing for a month or two and saves some honey for spring use.

In 1946 my bees had access to a small acreage of orange, not enough to produce a big yield but enough to provide typical orange conditions. So I kept the colonies small in winter when they had nothing coming in except a little from eucalyptus trees; and at the proper time I provided them with supplemental heat and fed them syrup and pollen mixture. They built up well, and worked on the orange without swarming, although other colonies in the vicinity did swarm. There would have been a big crop in my hives if the acreage had been sufficient.

16. THE WEATHER IS ALSO IMPORTANT. In northern California, similar conditions prevail in regard to apple bloom. When they have strong colonies and an abundance of stores, bees winter well, using a lot of honey, and can be ready for the early apple bloom. During my first year at Fort Bragg, the weather was good and my bees pro-

duced more apple-blossom honey than I had ever seen. They flew freely, pollinating the blossoms well and bringing about a very heavy set of fruit. We had a wonderful crop of honey and a wonderful crop of apples, pears, and plums.

The next spring, the bees were in as good or better condition; yet the apple-blossom honey harvested was almost nil; cold, windy weather prevailed during blossom time, and the bees flew hardly at all. When they did fly, they pollinated only the sunny side of a branch or tree. We had a very poor crop of fruit, of course, and very little honey.

Anyone can produce honey in seasons of plenty; but it takes a good beekeeper to make the best of poor years.

17. ROYAL JELLY. Royal jelly is the secretion from the pharyngeal glands of the nurse bees. It is so called in the belief that it was fed exclusively to queen larvae; but it is now quite well established that all larvae are fed it but that the queen larvae consume more than the others. It is enormously rich in vitamins, high in protein and fat, and low in sugar as compared with honey. Recently, some physicians have prescribed it for "run-down" patients, underdeveloped children, etc., with marvelous results. One doctor has been paying over $500 per pound for the product! (See *Gleanings in Bee Culture*, February 1954.)

18. SOME FACTS ABOUT HONEY. Honey, too, is regarded as having much therapeutic value. Generally, the darker honeys, although not so well liked for the table, have more vitamins and minerals than the lighter ones. Honey is favored by physicians and parents for sweetening children's food, modifying the baby's milk, etc. Honey spread on a burn gives quick relief. A tablespoon of honey with a tablespoon of lemon juice or vinegar in a large glass of water provides a quick pick-up when one is weary.

Honey will granulate more quickly in a cold place

than in a very warm one, and in melting it you should not heat it to more than 160°. Remember also that ripe honey keeps better than unripe honey does.

19. WHY DOES HONEY FERMENT? There are yeast spores in almost all types of honey, and where the humidity is high it is not easy for the bees to cure or ripen it. When the combs have been completely capped over, most of the excess moisture content has been removed. This is done by the bees' fanning. So having your hive well ventilated, with a good circulation of air, helps the bees to ripen it. When you are extracting your honey, avoid mixing well-capped combs with those that are not fully capped. Extract from the poorer combs last, after all the best honey has been removed from the extractor, and take care that the poorer honey is used up first.

If you have supers waiting to be extracted, store them in a dry, warm place to prevent the honey from absorbing moisture. Moisture is necessary for the yeasts to start fermentation. If you wish, you may heat your extracted honey in a double boiler to 160°, as this too aids in retarding fermentation as well as granulation.

20. THE OBSERVATION HIVE. One of the most frequently visited exhibits in the National Museum at Washington, D.C., is an observation hive with sides of glass. The single-comb observation hive, with comb-honey sections above the brood-nest comb, may be studied to advantage in the school or home or at an exhibition. Some beekeepers have made observation hives of standard size. Indoors, an observation hive may be placed several feet from a window, with a long tube connecting the hive to the outside of the house; the bees will learn to go through this tube.

With double thicknesses of glass and a quarter-inch space between, as in the Miller type of observation hive, the bees will live safely through the winter in the window of an unheated room. No shutters over the glass are required in this type of hive, as the bees soon become accustomed to the light. Hence their work may be observed at all times without disturbing them in the least. Such a hive may be purchased already made, but you might like to make one for yourself.

In the hive at the National Museum, the frames are shifted often enough so that, at all times, interesting work can be seen through the glass sides.

Some observation hives are made to take sections cut lengthwise through the cells, providing cells which have one side formed by the sheet of glass. Other cells are cut through crosswise, at the septum between the double row, so that their hexagonal bases can be seen. You can then see the bees produce wax scales and build comb. You also can see a queen back into a cell and deposit an egg, see the egg hatch, and observe nurse bees coming and

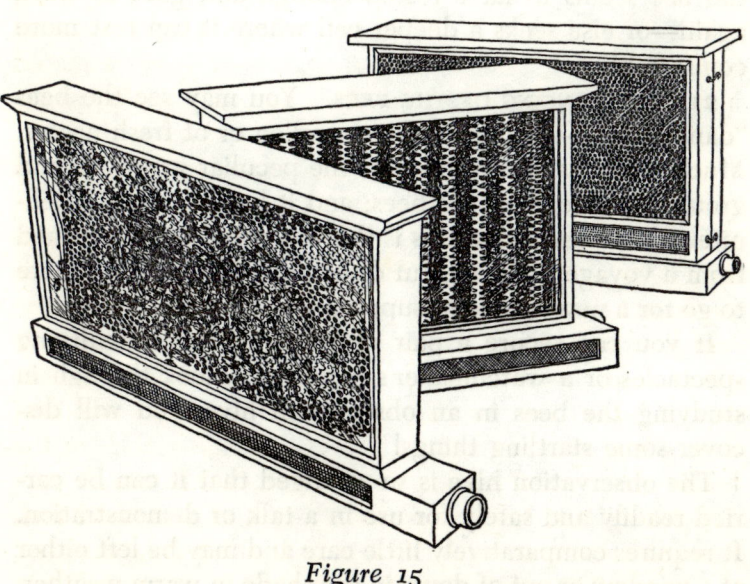

Figure 15

Observation Hives

feeding the larvae. You can watch the growth of a larva, see it spin its cocoon, change into a bee, and cut its way out of a cell.

Sometimes you will see a bee go into a cell head first and stay there some time. What is it doing? Perhaps it is cleaning and polishing the cell for the queen to lay in, or packing in pollen, or feeding a larva. Or it may be that it has made several trips to the field and has hidden away to rest awhile. I have seen a bee go into a shallow cell near the edge of the comb, unfinished, where it could conceal only half its length. At first the abdomen sticks straight out but after a while the tail begins to droop. Slowly it will go down until it hangs down too much for comfort (I suppose), then it will straighten up with a jerk—for all the world like the head of a man who has caught himself going to sleep in church! This may happen several times before the bee seems to have rested enough and goes to work again—or else seeks a deeper cell where it can rest more comfortably.

21. THE DANCING OF THE BEES. You may see the bees "dance," too—the dance of new pollen or of fresh nectar. Much study has been made of the peculiar gyrations and group movements of the bees; and it has been quite definitely determined that this is the way some bees, returned from a voyage of successful discovery, tell the rest where to go for a new source of supply.

If you can secure a pair of watchmaker's magnifying spectacles or a watchmaker's monocle to look through in studying the bees in an observation hive, you will discover some startling things!

The observation hive is so arranged that it can be carried readily and safely for use in a talk or demonstration. It requires comparatively little care and may be left either at a window or out of doors in the shade in warm weather.

22. **BE OPEN-MINDED TO NEW THINGS.** Remember the old saying:

> Be not the first by whom the new is tried,
> Nor yet the last to lay the old aside.

Young beekeepers—and you are a young beekeeper even though you may be past 60 if you are just a beginner—will be less afraid to try new things than some of those who are old in the industry. I think of new beekeepers as being the hope of progress in the beekeeping world. Keep in touch with others, and discuss your problems with them. There is really no topic in nature more interesting than bees.

5. *What Next?*

After you have finished this book, it will be wise to read it over again. At a second reading, you have the background of a wide view and will understand much better the how and the why of the earlier directions given. As you read it the second time, you can make your plans for equipment and location with a better appreciation of the problems which will confront you and of how to solve them. Moreover, this reading will give you confidence in yourself and your ability to meet varying situations. You will become a beekeeper with some assurance of success.

Appendix

BEE JOURNALS

STANDARD BOOKS ON BEEKEEPING

OTHER WORKS BY THE AUTHOR

Bee Journals

Following are the principal journals published in the English-speaking world. If one in another language is desired, write for the name and address to Division of Bee Culture and Biological Control, Beltsville, Maryland. (Occasionally one is advertised in an American journal.)

American Bee Journal, Hamilton, Ill. $2.00 a year; two years, $3.50; three years, $5.00.

Australasian Beekeeper, Box 20, P.O., Maitland, 3N, New South Wales, Australia. International Money Order 18 shillings.

British Bee Journal. $4.00 a year. Subscribe through the *American Bee Journal.*

Canadian Bee Journal, Streetsville, Ontario, Canada. $1.75 a year.

Gleanings in Bee Culture, The A. I. Root Co., Medina, O. $2.00 a year; two years, $3.50; three years, $5.00.

Modern Beekeeping, Clarkson, Kr. $1.50 a year; two years, $2.50; three years, $3.25.

Equipment supply houses will be advertised in all the journals and will forward catalogues upon request. Sample copies of the journals may also be requested.

Standard Reference Works and Books on Beekeeping

ABC and XYZ of Bee Culture, by E. R. Root, LL.D. A 720-page encyclopedia, alphabetically arranged. The A. I. Root Co., Medina, Ohio. $3.75.

The Hive and the Honey Bee, edited by Roy A. Grout. Dadant & Sons, Hamilton, Ill.

Allen Latham's Bee Book, Hale Publishing Co., Hapeville, Georgia. $2.95.

Honey Bees and Their Management, by Stanley B. Whitehead, Faber and Faber, 24 Russell Square, W.C. 1, London, England.

Starting Right with Bees. The A. I. Root Co. A fine handbook for the beginning commercial beekeeper. $1.00.

There are many other works, but these cover the ground very thoroughly. A list of government pamphlets on beekeeping may be had by writing to the Division of Bee Culture and Biological Control, Beltsville, Maryland.

Other Works by the Author

Honey Getting, 2nd Edition, 1947. $1.50.
The Beemaster System for Hive Heating, 1947. $1.00.
Scientific Beekeeping, by Sechrist and McFarland, 1948. $1.00.
The Twin Colony Hive. Plans and specifications. $2.00.

The publications above may be ordered from the publisher: Earthmaster Publications, Box 488, Sun Valley, Calif., or from Mrs. E. L. Sechrist, 1331 S. Juniper St., Escondido, Calif.

Index

Index

Abdomen, queen, 22, 59
Accessories, 6-7
Alighting board, 15
American foul brood, 61, 62
Angry bees, 10, 11, 14, 15
Antibiotics, 65, 124
Ants, 65, 66
Apiary, location, 14-16, 133, 134; shade, 15; windbreak, 15
Artificial fertiliation, 47

Balling of queen, 54
Basswood honey, 88
Bee candy, 54, 55, 109
Beebread, 42
Beekeepers' Associations, 128
Bees, clustering, 104; dancing, 138; diseases, 61-64, 72; escape, 4, 88, 89; guard, 35, 44; inspectors, 65, 128; journals, 4; number per colony, 41; races, 21-22; sleep, 138; space, 5, 26; venom, 18; weather, *See* Temperature
Birds, danger, 65
Bohne, George, 131
Brood, 29
Brood chamber, 29, 56
Brood nest, 29, 32, 56, 84; clear, 69, 72, 73, 82; inspection, 31, 32
Burr comb, 99

Cages, queen, 53-55; Shipping, *See* Package bees
Calcium cyanide, 62, 66
Cappings, 98
Carniolans, 21, 58
Caucasians, 21, 113, 114
Chantry principle, 55
Cheesecloth strainer, 92, 94
Chemicals, 62, 63, 128
Clipping wing, queen, 58, 59

Clothing, 9-14
Colony, balance, 73, 122; number of bees, 41; queenright, 71; strength, 74; two story, 30; *See also* Hive
Comb foundation, 4, 6, 7, 85
Comb honey, 85-87
Combs, brace, 34, 87; drone, 130; importance, 73, 83, 130
Corner cutting, 113
Cramps, of queen, 53

Dadant hive, 4, 6, 84; modified, 4, 6
DDT, 66
Denim, danger of, 9, 10
Demareeing, 59, 60
Dequeening, 55
Dextrose, 88
Diseases, 61-64
Dividing colonies, 75
Dragon flies, 65, 66
Drone, brood, 7, 29; cells, 130, comb, 130
Drone laying queens and worker, 60, 61
Drones, value, 45, 46
Dysentery, 124

Eggs, 41-42; laid by queen, 41-42, 61, 137; laid by workers, 60, 61
Enamel ware for cappings, 98
Enemies, 65, 66
Entrances, upper, 106; "warm way," 112, 113
Enzymes, 43
Equalizing chamber, 117
European foul brood, 61-65
Excluder, queen, 4, 7, 30, 56, 58, 60, 70, 85
Extractor, 93-95; solar wax, 100

Fanning by bees, 43, 136
Feeding bees, 109, 130
Fermentation of honey, 88, 136
Foundation, comb or brood, 6, 7
Frames, Hoffman, 27; moveable, 4-6, 27; replacing, 32; separating, 7, 8
Fruit damage, 132
Fuel for smokers, 8
Fumidil-B, 124

Gloves, 11, 12
Goldenrod honey, 88
Granulation, honey, 97
Guard bees, 35, 44

Heather honey, 97
Heating hives, 114-120
Heating honey, 93, 96, 97
Hive tool, 7, 36, 116, 119
Hives: Dadant, 4, 6, 84; double-walled, 105, 119; entrances, 106, 112, 113; Langstroth, 5, 6; standard, 4-6; twin colony, 131
Hoffman frames, 27
Honey, comb, 84-87, 90-91; density, 111; extracting, 91-93; fermentation, 136; good crop, 83, 131; granulation, 97; quantity, 131; ripe, 88; straining, 91; uncapping, 92; unripe, 88; weight, 88
Honey flow, preparing, 81
Honey Getting, 69, 70, 72
Honey stomach, 43
Horses, stung by bees, 15
Humidity in the hive, 106

Inspection of hives, 71-73; 86
Inspectors, county, 128
Insulation, 125
Italian bees, 21, 114

Langstroth hive, 5, 6
Larvae, 42, 135

Laying workers, 42, 60-61
Leveling of hives, 16, 85
Levulose, 88
Lice, 65
Life, length, 44

Mating, queen, 46-47, 51
Medicinal, honey, royal jelly, 135
Melting point, wax, 98
Mice, 65-66
Moisture in hive, 104-105
Mosquito hawks, 65-66

Nectar, composition, 88; evaporation, 43-44
Newspaper, 77-78, 127
Nosema apis, 65, 124
Nuclei, 71, 75, 76; uniting, 76-78
Nurse bees, 42-43, 137

Observation hive, 136-137
Onion honey, 88
Orientation flights, 43

Package bees, 121; installing, 123; time of hiving, 132
Packing for winter, 105-107
Parthenogenesis, 142
Pollen, 73, 127; baskets, 44; substitute, 125; trap, 125
Pollination, 78, 133
Propolis, 7, 45, 90-91; removal from hands, 45, 91

Queen bees, balling, 54; description, 41, 50-51; drone-laying, 60; failing, 41, 46; finding, 25-29; loss, 51-52; mating, 46-47
Queen excluder, 4, 7, 30, 56, 58, 60, 70, 85

Record keeping, 71
Requeening, 53-56

Ripening of honey, 88
Robbing, 30, 35, 96
Royal jelly, 135

Scientific Beekeeping, 43, 115
Sealing brood, 73; honey, 74, 89
Section honey, 87, 90
Separators, 87
Shade, necessity, 15
Skep, old-fashioned, 4
Skunks, 65
Smoke, 25-26, 31, 35
Smoker, 8-9, 23-25, 36
Southern Beekeeping, 74-75
Spiders, 65-66
Spring build-up, 127
Spring dwindling, 44, 126
Stings, avoiding, 10, 16-17; removing, 17-18
Sucrose, 88
Sugar syrup, 109, 120, 130
Sun on bees, 15, 122
Super covers, 107
Supers, 4
Supersedure, 48, 53, 125
Supplemental heat, 70-71; 74; 114-120
Swarms, 49, 57; hiving, 78-79; prevention, 57-59, 130

Temperature, 70-71, 103, 108; heating honey, 93; melting wax, 98-99; creaming honey, 97; postpone granulation, 91
Thermostat setting, 117-118

Thickness of honey, 96
Tools, 7
Twin colony hive, 131-132

Uncapping knife, 93-94
Uncapping outfit, 91-92
Uniformity of colonies, 70
Uniting colonies, 76-77
Unripe honey, 88

Veils, 33
Ventilation, 59, 106-107, 136
Virgin birth of drones, 42
Virgin queen, 50-51
Vitamins, 135

Warm way entrance, 112
Warming equipment, 115
Water for bees, 16
Wax, bleaching, 101; floor, 98; melting point, 98; rendering, 98-101
Weather, 111, 134-135
Windbreak, 15
Wintering bees, 102-120, 125
Wire imbedder, 6
Worker bees, 42; drone laying, 60-61; duties, 42-43; length of life, 44
Wrapping hives for winter, 106-107

Young bees, duties, 42-43

Zipper fasteners, 9-10

Officially Withdrawn from Minnesota Hobby Beekeeper's Association Library